Springer Tracts in Advanced Robotics

Volume 52

Andreas Nüchter

3D Robotic Mapping

The Simultaneous Localization and Mapping Problem with Six Degrees of Freedom

Springer

Professor Bruno Siciliano, Dipartimento di Informatica e Sistemistica, Università di Napoli Federico II, Via Claudio 21, 80125 Napoli, Italy, E-mail: siciliano@unina.it

Professor Oussama Khatib, Robotics Laboratory, Department of Computer Science, Stanford University, Stanford, CA 94305-9010, USA, E-mail: khatib@cs.stanford.edu

Professor Frans Groen, Department of Computer Science, Universiteit van Amsterdam, Kruislaan 403, 1098 SJ Amsterdam, The Netherlands, E-mail: groen@science.uva.nl

Author

Dr. Andreas Nüchter
University of Osnabrück
Institute of Computer Science
Albrechtstraße 28
Room: 31/506
49069 Osnabrück
Germany
E-Mail: Andreas@nuechti.de

ISBN 978-3-540-89883-2 e-ISBN 978-3-540-89884-9

DOI 10.1007/978-3-540-89884-9

Springer Tracts in Advanced Robotics ISSN 1610-7438

Library of Congress Control Number: 2008941001

Typeset & Cover Design: Scientific Publishing Services Pvt. Ltd., Chennai, India.

Printed in acid-free paper

5 4 3 2 1 0

springer.com

STAR (Springer Tracts in Advanced Robotics) has been promoted under the auspices of EURON (European Robotics Research Network)

To Jutta

Series Editor's Foreword

By the dawn of the new millennium, robotics has undergone a major transformation in scope and dimensions. This expansion has been brought about by the maturity of the field and the advances in its related technologies. From a largely dominant industrial focus, robotics has been rapidly expanding into the challenges of the human world. The new generation of robots is expected to safely and dependably co-habitat with humans in homes, workplaces, and communities, providing support in services, entertainment, education, healthcare, manufacturing, and assistance.

Beyond its impact on physical robots, the body of knowledge robotics has produced is revealing a much wider range of applications reaching across diverse research areas and scientific disciplines, such as: biomechanics, haptics, neurosciences, virtual simulation, animation, surgery, and sensor networks among others. In return, the challenges of the new emerging areas are proving an abundant source of stimulation and insights for the field of robotics. It is indeed at the intersection of disciplines that the most striking advances happen.

The goal of the series of Springer Tracts in Advanced Robotics (STAR) is to bring, in a timely fashion, the latest advances and developments in robotics on the basis of their significance and quality. It is our hope that the wider dissemination of research developments will stimulate more exchanges and collaborations among the research community and contribute to further advancement of this rapidly growing field.

The monograph written by Andreas Nüchter is yet another title in the series belonging to the fertile research area of self-localization and mapping (SLAM) for mobile robots. The main contribution of the work is certainly in the derivation of SLAM algorithms for building 3D maps, as opposed to the majority of existing algorithms for 2D maps. When tackling the problem for a 3D geometry, all six degrees of freedom are considered to reconstruct the robot pose. Results span from theory through a rich set of experiments, revealing a promising outlook toward benchmarking SLAM robot maps in a wide range of applications.

Expanded from the author's doctoral dissertation (finalist for the Seventh Edition of the EURON Georges Giralt PhD Award) and overall eight years of research and teaching on SLAM, this volume constitutes a very fine addition to STAR!

Naples, Italy,
August 2008

Bruno Siciliano
STAR Editor

Foreword

Have you ever jumped to bang on the emergency stop button of your mobile robot, to prevent it from crashing into an open drawer above its collision avoidance laser scan plane? That was the moment when you learned that the world is a 3D place. Well, maybe you guessed it before, which is the reason why you would painstakingly avoid to let the robot get close to staircases: You do know, but your robot may not know, that this wide open area so tempting in the 2D scan on which its heading direction control is based, is right above the flight of stairs down.

While it has always been possible to calculate 3D geometry from stereo images, recent years have witnessed creative usage of 2D laser scanners to get it – be it by having scanners rotate or pitch, or by mounting them at angles between the scan plane and the robot motion direction. Given the rocketing in performance of SLAM for 2D maps during the same period, it is only natural to think about building 3D geometry maps using these sensor data, so: How turn into a consistent 3D geometry map the point clouds generated by a 3D laser scanner on-board a mobile robot?

The obvious approach to do this is: Take your favorite SLAM algorithm for building 2D maps, say, a Particle Filter algorithm, and apply it to 3D rather than 2D points. That could work, but there is a problem. When thinking about 3D geometry, it makes sense to consider for the robot rigid body motion in 3D, i.e., in six degrees of freedom (6 DoF: x, y, z, yaw, pitch, roll). A typical SLAM algorithm for building 2D maps from 2D laser scans would consider only robot motion on a plane, i.e., 3 DoF poses. If it uses for building the map an approximation of the full posterior distribution of robot poses in pose space, then changing from 3 to 6 DoF poses lets explode the distribution. Put briefly, not all that works well in 3 DoF pose space is practical in 6 DoF.

Andreas Nüchter's book is a comprehensive introduction into approaches that do work well on 3D scan data in 6 DoF – in fact, it is the most comprehensive that I know of. It is based on Andreas's own research and teaching on the topic since about the year 2000. This has involved a number of different robots, and a healthy number of applications; it has earned him a diploma degree and a doctorate; it has yielded for his respective working groups grants and awards, such as a vice-championship in the RoboCup real rescue league (2004 in Lisbon), not to mention papers in all major robotics conferences and journals. So one can safely say, this author knows his topic well. Moreover, Andreas could exploit various teaching events for annealing the presentation.

The basic technique that underlies 6DoF SLAM here is the good old ICP (iterative closest/corresponding points) algorithm for registering 3D point clouds. If you don't know ICP yet, you will learn it from Chapter 4; read Chapter 4, too, if you ever wanted to know how to implement and use ICP really efficient, or how to choose between different versions of the math behind the required 6 DoF transformations or of the algorithms behind finding corresponding points. Then, scan registration is the basis of mapping here, but left alone would normally produce maps with unbounded errors. To build good maps, SLAM needs information about closed loops in the robot trajectory – this is no different for 6 DoF than it is for 3 DoF. Chapter 5 describes how this works in 6 DoF. The noblest technique here is a version of the Graph-SLAM algorithm, maximizing the joint probability of 6 DoF scan poses. So we are back at a probabilistic approach, yet a uni-modal one, for the good reason stated above.

The book also contains a healthy amount of results on data from real applications, where many of the data sets can be downloaded from an online 3D scan repository, to reproduce the results or do own experiments. Here, I think the topic of Benchmarking is particularly interesting in terms of methodology. You find many papers with SLAMmed robot maps that "look good"; you find few papers, at least if it comes to 3D maps, that question explicitly whether their maps are also truthful. In fact, that question is often hard to answer if the mapped area is thousands of square meters in open terrain with significant elevation differences. Section 6.4 tells you how it could be answered anyway. Finally, there is a section about Semantic Mapping in 3D: If benchmarking is a hot topic for applications, bringing semantics into robot-generated maps and into the mapping process is most probably the emerging hot topic for AI related robotics research.

So if you are interested in getting serious with 3D geometry in robotics, I recommend strongly that you read and understand this book. If you aren't interested in robotics, but only in 3D maps, there may be good reasons to read it anyway: The map building algorithms presented here don't really care whether their data is taken on-board a mobile robot or on a surveyor's tripod or on some hand-held scanning device. Give them 3D point clouds with rough 6 DoF pose estimates, and they allow consistent maps to be built without further manual interference. I am looking forward to seeing a lot of applications of that – in robotics and beyond.

Osnabrück, Germany Joachim Hertzberg
Juli 2008

Preface

Intelligent autonomous acting in unstructured environments requires 3D maps with labelled 3D objects. 3D maps are necessary to avoid collisions with complex obstacles and to self-localize in six degrees of freedom (x-, y-, z-position, roll, yaw and pitch angle). Meaning becomes inevitable, if the robot has to interact with its environment. The robot is then able to reason about the objects; its knowledge becomes inspectable and communicable. These arguments lead to the necessity of automatic and fast semantic environment modelling in robotics. A revolutionary method for gaging environments are 3D scanners, which enable robots to scan objects in a non-contact way in three dimensions. The presented work examines and evaluates the algorithms needed for automatic semantic 3D map building using a 3D laser range finder and the mobile robot Kurt3D.

The first part of this book deals with the task of registering 3D scans in a common coordinate system. Correct, globally consistent models result from a 6D SLAM algorithm. Hereby 6 degrees of freedom of the robot pose are considered, closed-loops are detected and the global error is minimized. 6D SLAM is based on a very fast ICP algorithm. While initially written for the mobile robot Kurt3D, the algorithms have also been applied in many other contextes. The second part presents a wide variety of working experiments and applications, including the results of presentations, like the Kurt3D presentation at RoboCup.

In the last part semantic descriptions are derived from the 3D point model. For that purpose 3D planes are detected and interpreted in the digitalized 3D scene. After that an efficient algorithm detects objects and estimates their pose with 6 degrees of freedom.

Of course, this book is the result of my active cooperation with many researchers. I would like to express my thanks to my supervisor Joachim Hertzberg who has always supported me. I am also thankful for the joint preceding research on 3D robotic mapping with: Dorit Borrmann, Thomas Christaller, Jan Elseberg, Simone Frintrop, Dominik Giel, Matthias Hennig, Joachim Hertzberg, Achim Lilienthal, Kai Lingemann, Martin Magnusson,

Stefan Stiene, Hartmut Surmann, Sebastian Thrun, Bernardo Wagner, Oliver Wulf. Special thanks go to Alexander Loktyuhsin for proofreading the book.

My intention is that the reader of this book will use this compilation of algorithms and results to build smarter robotic systems that can cope with real world environments using 3D sensor data.

Osnabrück, Germany Andreas Nüchter
Juli 2008

Contents

1 Introduction 1
1.1 A Brief Introduction to Robotic Mapping 2
1.2 The Big Picture 7

2 Perceiving 3D Range Data 9
2.1 Basic Principles 9
2.1.1 Time-of-Flight Measurement 9
2.1.2 Phase-Shift Measurement 10
2.1.3 Triangulation 11
2.2 Laser Scanner 11
2.2.1 1D Laser Measurement System 11
2.2.2 2D Laser Scanner 11
2.2.3 3D Laser Scanner 12
2.2.4 Projection 3D Scanner 14
2.3 Cameras and Camera Models 15
2.3.1 The Pinhole Camera Model and Perspective Projection 16
2.3.2 Stereo Cameras 19
2.4 3D Cameras 26

3 State of the Art 29
3.1 Categorization of Robotic Mapping Algorithms 29
3.1.1 Planar 2D Mapping 29
3.1.2 Planar 3D Mapping 31
3.1.3 Slice-Wise 6D SLAM 32
3.1.4 Full 6D SLAM 32
3.1.5 Recent Trends 32
3.2 Globally Consistent Range Image Alignment 33

4 3D Range Image Registration ... 35
4.1 The ICP Algorithm ... 35
4.1.1 Direct Solutions of the ICP Error Function ... 36
4.1.2 Approximate Solution of the ICP Error Function by a Helical Motion ... 48
4.1.3 Linearized Solution of the ICP Error Function ... 50
4.1.4 Computing Closest Points ... 52
4.1.5 Implementation Issues ... 64
4.1.6 The Parallel ICP Algorithm ... 67
4.2 Evaluation of the ICP Algorithm ... 69

5 Globally Consistent Range Image Registration ... 77
5.1 Explicit Loop Closing ... 77
5.1.1 Extrapolating the Odometry to Six Degrees of Freedom ... 77
5.1.2 Closed Loop Detection ... 78
5.1.3 Error Diffusion ... 79
5.2 GraphSLAM as Generalization of Loop Closing ... 81
5.2.1 Algorithm Overview ... 81
5.2.2 Probabilistic Global Relaxation Based on Scan Matching ... 81
5.2.3 Transforming the Solution ... 87
5.3 Other Modeling Approaches ... 93
5.3.1 Linearization Using Quaternions ... 93
5.3.2 GraphSLAM Using Helical Motion ... 99
5.3.3 GraphSLAM Using Linearization of a Rotation ... 104

6 Experiments and Results ... 109
6.1 3D Mapping with Kurt3D ... 109
6.1.1 The 3D Laser Range Finder ... 109
6.1.2 The Mobile Robot Kurt3D ... 110
6.1.3 Full 6D SLAM in a Planar Environment ... 111
6.1.4 Full 6D SLAM in an Indoor/Outdoor Environment ... 112
6.1.5 Full 6D SLAM in an Outdoor Environment ... 114
6.1.6 Kurt3D @ Competitions ... 121
6.2 Globally Consistent Registration of High Resolution Scans ... 123
6.3 Mapping Urban Environments – Registration with Dynamic Network Construction ... 125
6.4 Benchmarking 6D SLAM ... 130
6.4.1 Ground Truth Experiments ... 133
6.4.2 The Benchmarking Technique ... 134
6.4.3 Experimental Results ... 138
6.4.4 Justification of the Results ... 141
6.5 Further Applications ... 145
6.5.1 Mapping Abandoned Mines ... 145
6.5.2 Applications in Medicine ... 151

7 3D Maps with Semantics 155
7.1 Scene Interpretation 155
7.1.1 Feature Extraction 155
7.1.2 Interpretation 156
7.1.3 Application: Model Refinement 160
7.2 Object Localization in 3D Data 163
7.2.1 Object Detection 163
7.2.2 Object Localization 169
7.3 Semantic 3D Maps 170

8 Conclusions and Outlook 173

A Math Appendix 177
A.1 Representing Rotations 177
A.1.1 Euler Angles 177
A.1.2 Axis Angle 178
A.1.3 Unit Quaternions 179
A.1.4 Converting Rotation Representations 181
A.2 Plücker Coordinates 182
A.3 Additional Theorems from Linear Algebra 183
A.4 Solving Linear Systems of Equations 186
A.5 Computing Constrained Extrema 187
A.5.1 Computing the Minimum of a Quadratic Form with Subject to Conditions 188
A.6 Numerical Function Minimization 190

References 193

Notation

$\mathbb{R}$	set of reel numbers
$\mathbb{1}$	identity matrix
M, D	set of measured data, subset of $\mathbb{R}^3$
$x, y, z, \ldots$	scalars
$\boldsymbol{x}, \boldsymbol{y}, \boldsymbol{z}, \ldots$	vectors
$\dot{\boldsymbol{q}}, \dot{\boldsymbol{s}}, \ldots$	quaternions
$\boldsymbol{R}, \boldsymbol{H}, \ldots$	matrices, including rotation matrices
$\mathbf{M}, \mathbf{G}, \ldots$	matrices and vectors that have been concatenated
$\mathrm{tr}\,(\boldsymbol{M})$	trace of the matrix $\boldsymbol{M}$

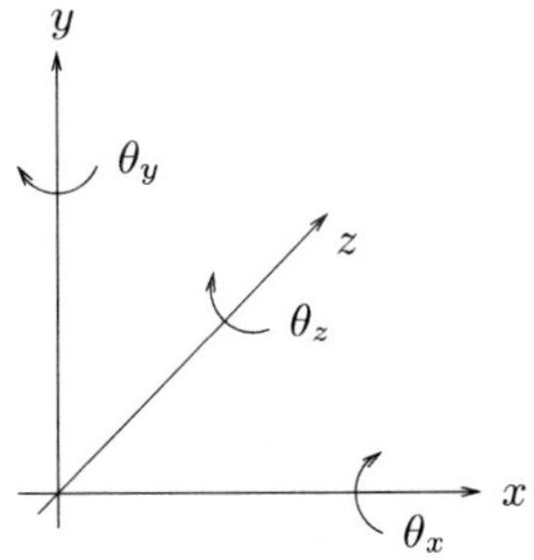

Definition of the coordinate system.

Chapter 1
Introduction

Humans have always dreamed of building artificial beings that take over tedious, annoying or dangerous tasks, that entertain them and that are subject to human command. This fascination is reflected in literature. Golems made of clay, Homunkuli, i.e., bio-chemical creatures, or androids, that are pure technical devices come into play in ancient times [131].

Even in the 50th, when the term "electronic brain" was established, the belief of many humans was that computers will think just like humans very soon and that they will become smarter some day. At this time optimistic computer experts established the new computer science discipline "artificial intelligence" (AI) aiming at teaching computers to talk, to understand spoken language, to translate languages, to give advices, to recognize images and handwriting, to control robots on far planets or to play chess. Today the past optimism seems to be exaggerated, since many computer programs and the used AI do not convince.

For technical devices it is among other things difficult to perceive the environment. In defined environments computer vision is able to check if for example bottles are correctly cleaned, if a work piece has the correct predefined form and pose, or if a solder joint is correct. This is basically done by cameras and fast image comparison.

However, general environments require much more for robot vision: The chaotic sensor data including lines, shadows, light reflexes must be analyzed and interpreted. If a robot should interact in such an environment in a goal-directed way the robot must build an internal representation, i.e., map and detect objects reliably. In natural environments it is necessary that the environment is perceived in 3D, which includes depth. Otherwise objects with jutting out edges cause problems. In addition complex robotic tasks need interpreted 3D maps, i.e., maps with interesting objects being labeled. In this way the knowledge of the robot is communicable and inspectable.

This book presents algorithmical and technical methods for mapping environments in 3D using a 3D laser scanner. The so-called simultaneous

A. Nüchter: 3D Robotic Map.: The Simultaneous Local. & Map. Prob., STAR 52, pp. 1–7.
springerlink.com

Fig. 1.1 The mobile robot Kurt3D. Photo: Dieter Klein.

localization and mapping problem is solved for 3D maps and robot poses with six degrees of freedom (6 DoF), hence 6D simultaneous localization and mapping (6D SLAM). The methods have been developed for the mobile robot Kurt3D (see Figure 1.1), but have also been tested on other robots and in other setings.

1.1 A Brief Introduction to Robotic Mapping

Solving the simultaneous localization and mapping problem (SLAM) is one of the important questions in robotics and it has attracted recently a lot of attention [26, 38, 78, 122, 124]. To map an environment state of the art systems use probabilistic methods and landmarks. They regard SLAM as an estimation problem. During robot localization the state of the robot $\boldsymbol{x}_t$, e.g, the pose of the robot, has to be estimated from all previous sensor readings $\boldsymbol{z}_{1,\ldots,t-1}$ and previous robot actions $\boldsymbol{u}_{1,\ldots,t-1}$. Robot localization is often solved using the Markov assumption, i.e., the world is assumed to be static, noise to be independent and no approximation errors during modeling to be made. Using these Markov assumptions the estimation problem for localization simplifies to the Bayes net given in Figure 1.2.

Mapping extends the estimation problem. In addition to the robot state the map $\boldsymbol{m}$ has to be estimated, as given in Figure 1.3. Usually, the map consists of n landmarks. In this case, the position of these landmarks have to be estimated to perform mapping.

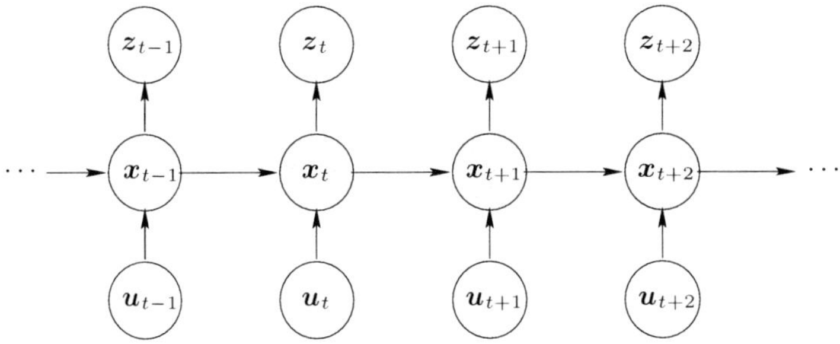

Fig. 1.2 Bayes net for robot localization

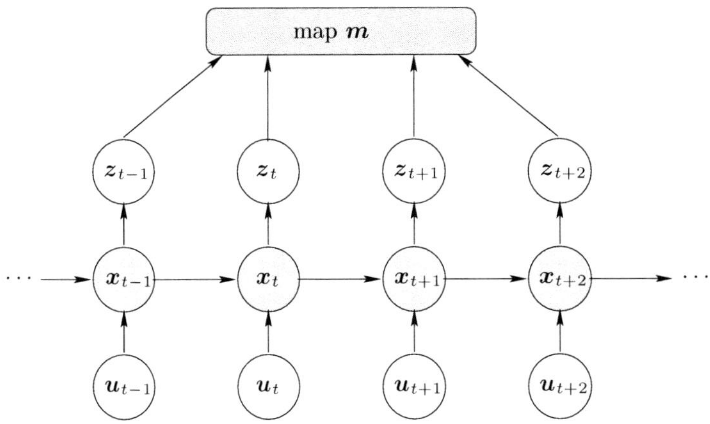

Fig. 1.3 Bayes net for robotic mapping

If a robot moves through a known environment, that is, a map is given to the robot, the uncertainty of the robot pose can be bounded. The observation of landmarks allows the precise estimation of the robot pose up to the uncertainty of the landmark position and observation. However, if the robot moves through an unknown environment, the uncertainties of the robot poses can be arbitrarily large: Odometry errors sum up. Tracked landmarks provide a basis for reducing the uncertainty in the robot pose. For many sensors such a tracking outperforms pure odometry [75,121]. But this does not change the principle of error accumulation. This circumstance makes the SLAM problem different from other estimation problems. Sensors working in a global frame of reference like a compass or GPS are a workaround and not always available, e.g., within buildings.

Closed loops play an important role in known SLAM algorithms. If a robot detects a position, where it has been before and correctly matches landmarks, then the accumulated error can be bounded. SLAM algorithms deform the map to compute a topologically consistent map. The following probabilistic methods have been used to solve the SLAM problem:

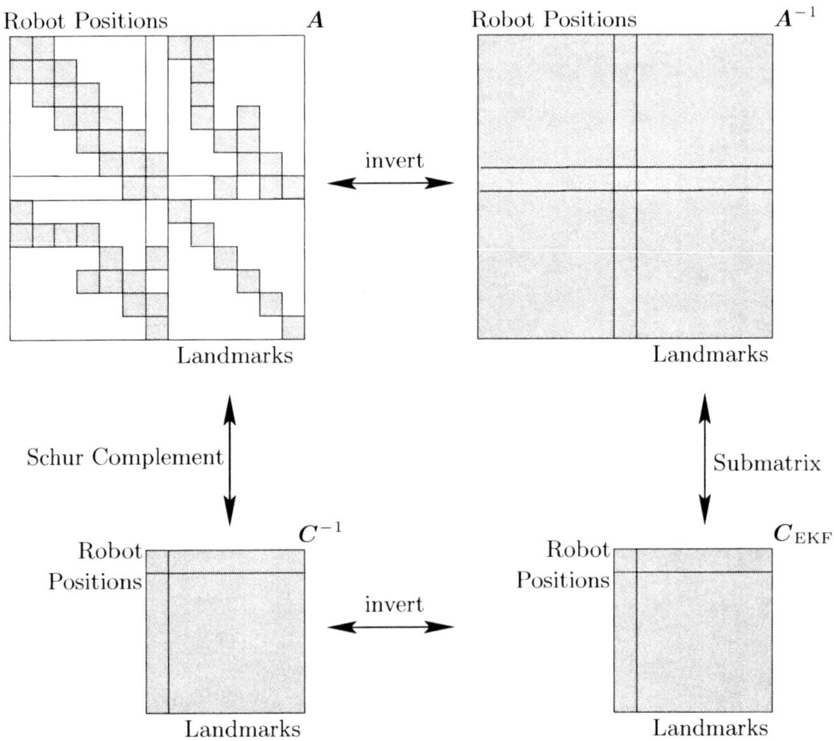

Fig. 1.4 Dependency of the information matrix $\boldsymbol{A}$ and the covariance matrix $\boldsymbol{C}_{\mathrm{EKF}}$ of the extended Kalman filter. Based on [38].

Maximum Likelihood Estimation. If one assumes independent Gaussian distributed measurement errors with known covariance, the SLAM problem can be reduced to minimizing a quadratic error function $Q(\boldsymbol{x})$ (non-linear model fitting) [38]:

$$Q(\boldsymbol{x}) = \sum_i \frac{1}{2} (\boldsymbol{y}_i - \boldsymbol{f}_i(\boldsymbol{x}))^T \boldsymbol{C}_i^{-1} (\boldsymbol{y}_i - \boldsymbol{f}_i(\boldsymbol{x})) \tag{1.1}$$

$$\boldsymbol{x}_{\max} = \arg\min_{\boldsymbol{x}} Q(\boldsymbol{x}).$$

Here $S_\gamma := \{\boldsymbol{x} | Q(\boldsymbol{x}) - Q(\boldsymbol{x}_{\max}) < \gamma\}$ is the region that encloses $\boldsymbol{x}_{\max}$ and the parameter γ defines the uncertainty in the map. The poses of the robot are concatenated to the vector $\boldsymbol{x}$. $\boldsymbol{y}_i$ is the i-th measurement with the covariance matrix $\boldsymbol{C}_i$ and $\boldsymbol{f}_i(\boldsymbol{x})$ is the corresponding measurement function. This measurement function returns the value, that the measurement should have if the landmark is visible from pose $\boldsymbol{x}$.

Minimizing Eq. (1.1) means finding a good map. It can be performed using the Levenberg-Marquardt algorithm. An incremental SLAM method using this method with p robot poses and n landmarks needs $\mathcal{O}((n+p)^3)$ computing time.

Extended Kalman Filter. Most often the extended Kalman filter is used to solve the SLAM problem. The filter integrates all measurements of landmarks in a covariance matrix of landmarks positions and robot poses. For linear measurements models this corresponds to the maximum likelihood method discussed above, for non-linear measurement models this only holds true, iff we can linearize the measurement function. Linearization is usually done while measuring [124], but the point for linearization can be significantly erroneous, if the robot moves through an unmapped area [38]. Here the error corresponding to the orientation is critical, since trigonometric functions have only a small quasi-linear interval. The result of erroneous linearization is, that closed loops are usually too large, in comparison with ground truth [38].

Information Filter. Linearization of the measurement function $\boldsymbol{f}_i$ yields a quadratic function Q:

$$Q(\boldsymbol{x}) = \boldsymbol{x}^T \boldsymbol{A}\boldsymbol{x} + \boldsymbol{x}^T \boldsymbol{b} + c.$$

The matrix $\boldsymbol{A}$ is the information matrix. Here we linearize again at the time of the measurement. $\boldsymbol{A}$ is a sparse matrix, i.e., contains many zero entries. Only the entries that correlate measurements are used. The inverse matrix $\boldsymbol{A}^{-1}$ is the complete covariance matrix for all robot poses and landmarks. Therefore, $\boldsymbol{A}^{-1}$ represents indirect dependencies and a submatrix is the covariance matrix $\boldsymbol{C}$ that is used in the extended Kalman filter. Figure 1.4 shows the dependencies between the information matrix and the covariance matrix.

Sparse Extended Information Filters (SEIFs) are used to process information matrices. To ensure linear computation time, the corresponding algorithms process the data such that the matrices are sparse [124].

A key challenge of all SLAM algorithms is data association that has to solve the question which features correspond to each other. The solution presented

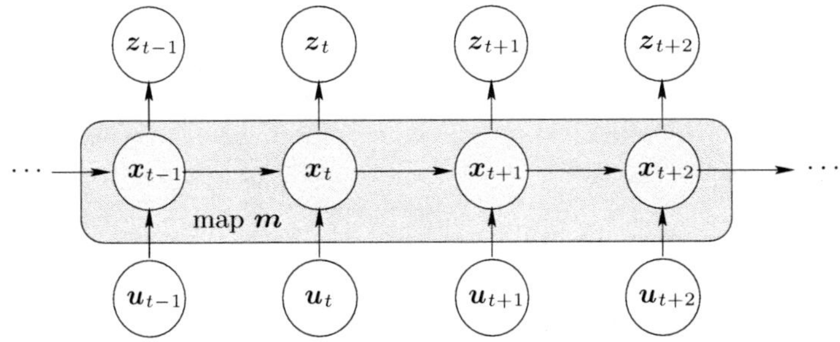

Fig. 1.5 Bayes net for robotic mapping used in this book

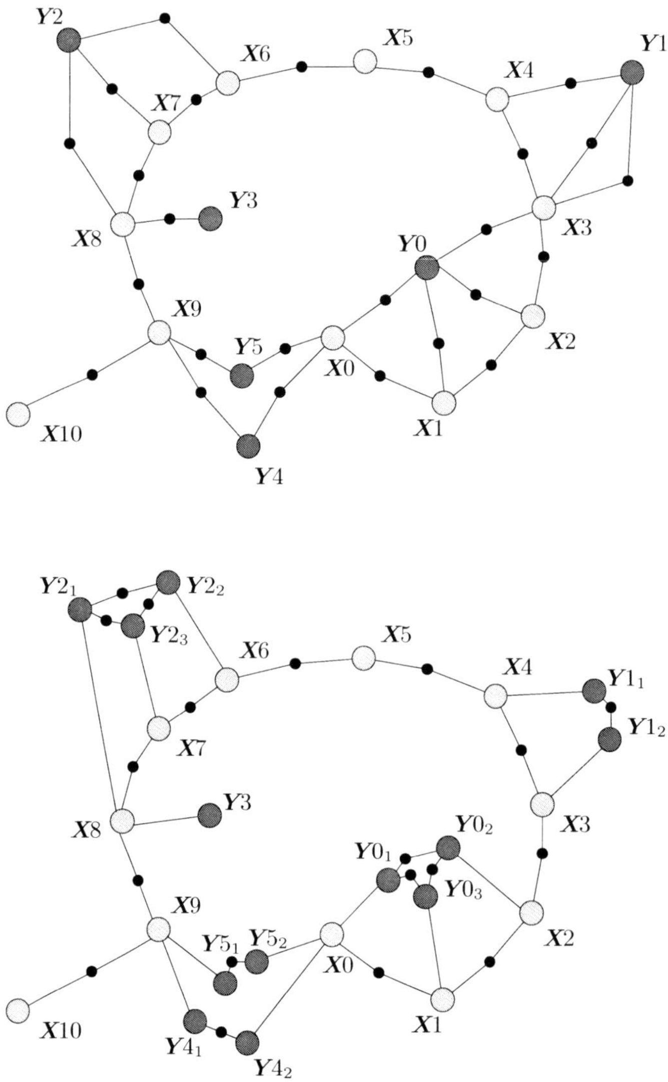

Fig. 1.6 Abstract comparison of SLAM approaches. Left: Probabilistic methods. The robot poses $\boldsymbol{X}_i$ as well as the positions of the associated landmarks $\boldsymbol{Y}_i$ are given in terms of a probability distribution. Global optimization tries to relax the model, where the landmarks are fixed. Small black dots on lines mark adjustable distances. Right: Our method with absolute measurements $\boldsymbol{Y}_i$ (note there are no black dots between scan poses and scanned landmarks). The poses $\boldsymbol{X}_i$ are adjusted based on scan matching aiming at collapsing the landmark copies $\boldsymbol{Y}_{i_k}$ for all landmarks $\boldsymbol{Y}_i$. Data association is performed by closest point search.

in this book uses nearest neighbor relations and avoids this problem. Through iterative computation of closest points the algorithms assume that in the last iteration the correspondences are correct.

1.2 The Big Picture

The SLAM algorithms mentioned so far require feature extraction. The methods presented in this book do *not* use features and work with plain 3D scans. This approach is similar to the methods developed by Lu and Milios [77,78]. SLAM is based on scan comparison. With every robot pose a scan is bundled. If two scans overlap a relation of the corresponding robot poses is built based on scan matching. The resulting graph of poses can be treated like a landmark-based graph. Since the positions of the landmarks do not have to be estimated, the problem is simplified.

Figure 1.5 presents our understanding of scan matching based SLAM. While in Figure 1.3 both robot poses *and* landmark positions have to be estimated, only robot poses are estimated here. When all poses have been estimated, the map is automatically given through rigid transformations of the sensor values based on the estimated poses.

A comparison of probabilistic SLAM approaches with our approach on an abstract level as presented by Folkesson and Christensen [34] is given in Figure 1.6. Robot poses are labeled with $\boldsymbol{X}_i$ whereas the landmarks are $\boldsymbol{Y}_i$. Lines with black dots correspond to adjustable connections, e.g., springs, that can be relaxed by the algorithms. In our system, the measurements are fixed and data association is repeatedly done using nearest neighbor search.

Probabilistic SLAM algorithms representing the robot poses with three degrees of freedom were very successful recently. The corresponding probability density functions have been represented/approximated by particles. As a consequence, stability has been reached through tracking multiple hypotheses. These probabilistic SLAM algorithms do not have been extended to generate 3D maps and regarding robot poses with 6 DoF. 3D maps generated by an additional vertically mounted 2D scanner regard poses with 3 DoF and are mainly used for evaluation of 2D mapping approaches [42,43,122].

The book presents a set of methods and algorithms for solving 6D SLAM. The outcome is a globally consistent registered 3D point cloud. In many examples we emphasize the applicability of the methods. The last chapter introduces semantics, aiming at interpreting 3D laser scans. A label is attached to the points, such that the points are categorized and classified.

Chapter 2
Perceiving 3D Range Data

Three dimensional images may be perceived in three different ways: Triangulation of corresponding points with known baseline results in a depth estimate. Another depth measuring method is available via the time-of-flight principle. As a third method depth may be acquired from a single image using background knowledge. In this chapter we discuss the first two methods and show their advantages and disadvantages. Single image approaches are currently not used in mobile robotics.

2.1 Basic Principles

2.1.1 Time-of-Flight Measurement

Time-of-flight measurements consists of measuring the delay between a emited signal returns to the receiver. Using the measured time t and the velocity v of the signal, the travelled distance is given by

$$s = v\, t.$$

Usually, the receiver is attached close to the sender and therefore we can assume that the distance sender/object equals the distance object/emitter (see Figure 2.1). Therefore, the distance to the object is

$$D = v\, \frac{t}{2}.$$

The velocity v for sound in air is $v = 343\,\mathrm{m/s}$, for light in vacuum $v = c = 299792458\,\mathrm{m/s}$.

Due to the high value of the speed of light a precise timer is needed. To measure a distance difference of 1 cm the precision of the time measurement must be in the range of pico-seconds:

A. Nüchter: 3D Robotic Map.: The Simultaneous Local. & Map. Prob., STAR 52, pp. 9–27.
springerlink.com

Fig. 2.1 Principle of active time-of-flight measurements. Signals that are emitted from the sender and reflected from the objects are detected by the receiver.

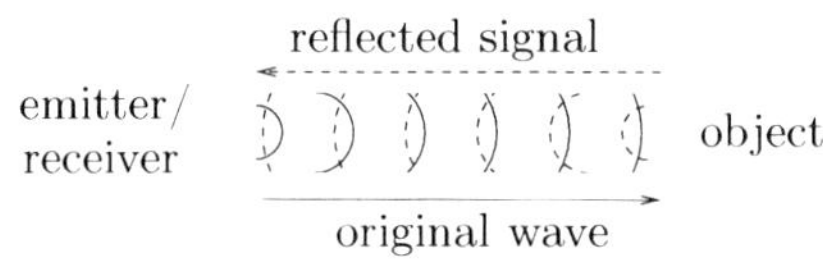

$$\Delta t = \frac{\Delta s}{c} = \frac{0.010\,\text{mm}}{299792458\,\text{m/s}} = 33.4 \times 10^{-12}\,\text{s}$$

2.1.2 *Phase-Shift Measurement*

Measuring phase shift also yields distances and is a special from of time-of-flight measurements. A continuous wave, e.g., a continuous wave of laser light, is emitted. The amplitude of the carrier signal is modulated by sinus-signals with different frequencies yielding the so-called measuring signals. The reflected signal is compared to the currently sent one and the phase shift $\Delta\varphi$ is measured. The phase shift is proportional to the distance D [12], i.e.,

$$D = \frac{\Delta\varphi\,\lambda}{4\pi} = \frac{\Delta\varphi\,v}{4\pi f}$$

where λ is the wave length of the modulated signal, f the frequency respectively. The carrier signal is not analyzed, since the wave length is very small and the distances can only be measured relatively to the signal waves.

It is not possible to determine the overall distance, since the phase shift is a relative proportion with respect to the wave length. Therefore, most phase shift devices use two modulation frequencies, a high frequency for high resolutions and a low frequency that measures large distances. Figure 2.2 shows the principle: A measuring signal is modulated with two other signals.

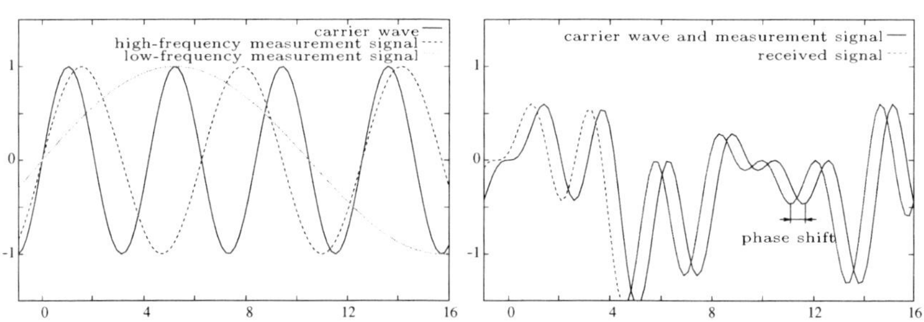

Fig. 2.2 Modulation of amplitudes. Left: carrier signal, low-frequency and high-frequency measurement signal (dashed). Right: Superposition of the carrier signal and the measurement signals and the corresponding reflected wave.

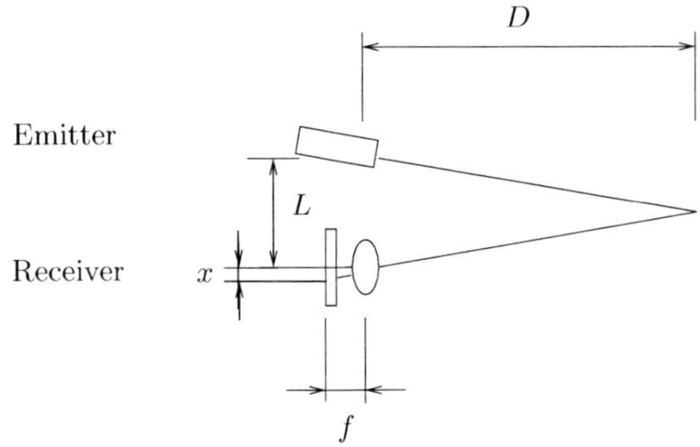

Fig. 2.3 Triangulation principle

2.1.3 Triangulation

Distances can be measured via triangulation. The sender emits a signal and at the receiver that is now attached at the distance L from the sender, where the signal is detected under a deviation of x using a pinhole camera. Based on the deviation x the distance is calculated, by

$$D = f\frac{L}{x} \tag{2.1}$$

The pinhole camera model permits the calculation above, since the theorem of intersecting lines can be applied. Figure 2.3 shows the triangulation principle.

2.2 Laser Scanner

In the last decade laser scanners emerged as a novel technology. They consist of a distance measuring system that uses a laser. A scanning method is used to digitalize surfaces, which requires a mechanics to move or rotate the measuring system. Alternatively, rotating mirrors are used to change the direction of the laser beam. The actual distance determination is done by a time-of-flight, phase shift or triangulation measurement.

2.2.1 1D Laser Measurement System

A Laser Measurement System (LMS) is a sensor, that uses laser light to measure distances. Using a serial port or Bluetooth the distance value is directly transmitted from an embedded system that calculates the distance to the host computer.

2.2.2 2D Laser Scanner

2D laser scanners extend 1D systems. To scan a line, the laser is rotated, or more often a mirror is rotated, such that the laser beam measures distances

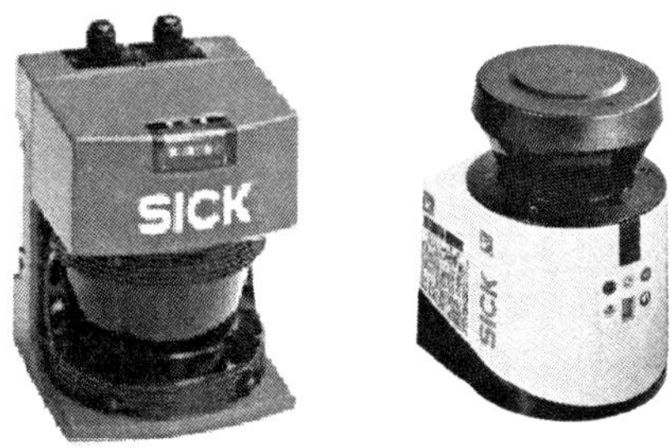

Fig. 2.4 SICK laser scanner (LMS200, S300)

in different directions. The apex angle, usually 90°, 180°, 270° or 360°, is discretized with different resolutions, e.g., 1°, 0.5° or 0.25°. Figure 2.4 shows two scanners that are commonly used in robotics. These scanners are safety scanners, which implies that they have a large beam divergence, such that *all* objects within the apex angle are detected.

Due to the beam divergence objects of the environment are hit by one laser shot. In this case the measured value can be the distance to the closest object, to the furthest object, or all values in between. Such edges are called jump edges and are often subject to erroneous values.

2.2.3 3D Laser Scanner

3D laser scanners extend 2D systems by an additional degree of freedom. Either a second mirror or the whole 2D scanner is rotated. Table 2.1 shows the possible configurations of a 2D scanner for acquiring 3D data. Depending on the configuration different advantages and disadvantages are yield. Figure 2.5 shows two 3D point clouds that are the result of different scanning methods. The different point densities as well as the scan slices can be seen. The latter ones are perceivable as lines.

Besides the safety scanners, 3D laser scanners are the state of the art tool in geodesic applications. In this context they are called LiDAR (engl.: *Light*

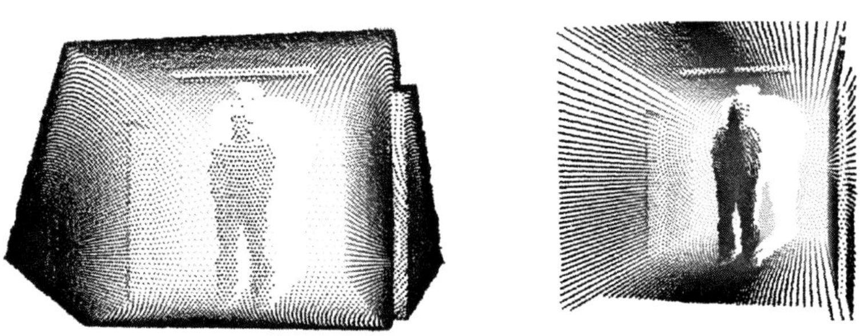

Fig. 2.5 Left: 3D point cloud scanned with a pitching scanner. Right: 3D point cloud scanned with a yawing scanner. Figures courtesy of Oliver Wulf, Leibniz Universität Hannover, Germany.

Table 2.1 3D scanner configurations using a SICK scanner. Figures courtesy of Oliver Wulf, Leibniz Universität Hannover, Germany.

Mode	Symbol	continuously rotating	tilting	Advantages and Disadvantages
Yaw				+ complete 360° scans – high point densities on top and bottom + good point arrangements using a scanner with 90° apex angle
Yaw-Top				+ fast scan time, since half a rotation is sufficient + high point density in viewing direction of the scanner – only a half-space scanable
Roll				+ fast scan time, since half a rotation is sufficient + high point density in viewing direction of the scanner – only a half-space scanable
Pitch				– high point density at the sides – only small fractions of the space scanable

Detection and Ranging) or LADAR (engl.: *Laser Detection and Ranging*). Geodesy has developed Terestrical Laser Scanning (TLS) as an independent field. All environments and objects are digitalized in 3D. The scanners also rotate a 2D scanning device that in turn extends a 1D measuring unit. Scanners that use the phase shift principle yield high accuracies and high data rates, e.g., the IMAGER 5006 of the company Zoller+Fröhlich scans up to 500000 points per second. Measuring systems based on pulsed laser technology are able to measure large distances, e.g., the Riegl scanner LMS-Z420i

Fig. 2.6 Left: 3D scanner produced by the company RIEGL Laser Measurement Systems GmbH [104]. Right: Top 2D view of a scan. The scan points have been colored using a calibrated camera. Right bottom: 3D view of the scene. Figures courtesy of Riegl.

detects Objects up to a distance of 1000 m. Figure 2.6 shows a geodetic 3D laser scanner and the corresponding scan.

The high accuracy of such LiDAR systems leads to high costs and therefore these scanners are seldom used in robotics. However, a recent trend in surveying is so-called kinematic laser scanning (k-TLS). The scanner is mounted on a mobile platform, e.g., a car or lane, and used in the profiler mode, i.e., in a 2D scanning mode. The additional dimension needed for acquiring 3D data is generated by the moving platform. The environment is digitalized with profiles. The accuracy of the 3D data depends on the accuracy of the positioning information (cf. Section 3.1.3).

2.2.4 Projection 3D Scanner

Projection laser scanners use the triangulation principle. A pattern is projected onto a scene and the image is captured by a camera. If the pattern is a laser spot, we have a 1D laser measurement system. If it is a line, 2D data can be extracted from the image. A planar pattern yields a 3D laser scanner. Figure 2.7 demonstrates the principle.

Computation of depth is done by the triangulation formula (2.1). This requires the calibration of the setup, i.e., the intrinsic and extrinsic parameters of the camera with respect to the light source must be known (see Section 2.3). Alternatively the depth for the 2D case may be computed based on the known angle α from the disparity x in the camera image:

$$H = x \ \tan(\alpha),$$

If the pose of the camera with respect to the light source is unknown, additional information must be used to compute the plane of the laser light. This can be achieved by placing the object in front of a defined corner that is also captured by the camera [135] (cf. Figure 2.7, right) or by using two laser lines orthogonally projected onto the scene [68].

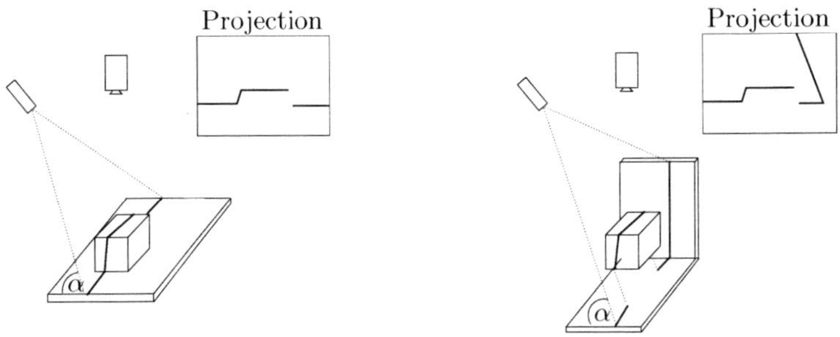

Fig. 2.7 Left: 3D information can be computed based on a projection of a line with the known angle α. Right: A predefined corner helps estimating the depth, in case α is unknown.

3D scanner project a 2D pattern. Cost effective versions use a SVGA-projector and binary patterns with horizontal and vertical lines with increasing number of lines. For every projected pattern the camera acquires a picture. Every point gets a unique binary code, i.e., grey code, that is detected in the image and therefore, the correspondences of camera pixels and projector pixel are given. Based on these correspondences depth images are computed similarly to stereo vision [106].

2.3 Cameras and Camera Models

Image acquisition is performed in two steps. First, the image is brought into focus. Cameras with autofocus use a rough distance measurement and adjust the lenses, otherwise manual adjustment is necessary. Second, the image sensor captures light using a defined timing. The image sensor produces then electrical signals from the light intensities. Afterwards an analog-digital converter quantifies the signal, that is transmitted to the computer.

CCD image sensors consist of a matrix of light sensitive pixels. Similarly to a photo diode a CCD sensor consists of a semiconductor that separates electrons and positive charges, i.e., holes, while exposed to light. The extension to a photo diode consists of collecting the charges and transferring them step-by-step towards an amplifier and an analog-digital converter. CMOS image sensors do not transfer the charges to a single amplifier and converter; here an amplifier is attached to every sensor. It seems that CMOS technology is more complex, but the production costs are lower, since more common semiconductor materials can be used. Furthermore, additional functionality, e.g., contrast correction or exposure control, can be integrated directly to the chip. A disadvantage of CMOS is that the sensing area is smaller and this technology is not so light sensitive. Webcams usually use CMOS, while digital cameras use a CCD chip.

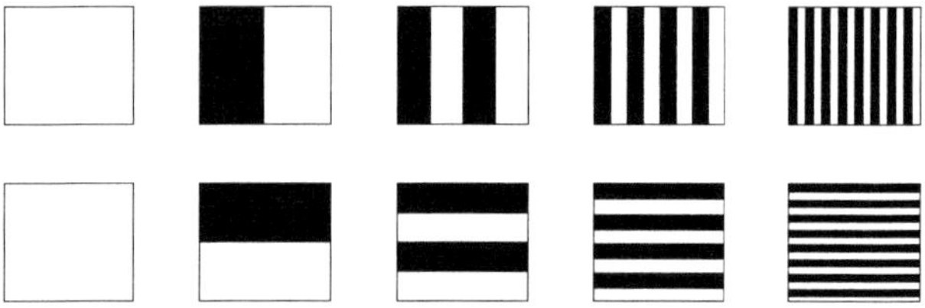

Fig. 2.8 Binary code for projection laser scanner

2.3.1 The Pinhole Camera Model and Perspective Projection

The projection in a camera can be described as a first approximation by the pinhole camera model. If light falls through a small hole into a box then on the opposite side of the hole a perspective projection occurs. The perspective projection of parallel lines yields intersections. Figure 2.9 shows an example of one vanishing point. The projection follows the law of intersecting lines, as given in Figure 2.10. All rays go through the focal point $\boldsymbol{o}$. Objects of same size that are at the same distance to $\boldsymbol{o}$ yield projections of the same size.

A point $\boldsymbol{p}$ in the three-dimensional space has the coordinates $({}^cx, {}^cy, {}^cz)$ relative to the camera center $\mathbf{c}$. The point $\boldsymbol{p}$ is projected to the image point $\boldsymbol{p}'$ in the image plane with the coordinates (u, v). The latter coordinates are relative to the image center $\mathbf{c}'$ that is the intersection between the optical axis and the image plane (cf. Figure 2.11). The dependence of the 3D coordinates in the camera coordinate system and the image coordinates with a focal length f is given by:

$$-\frac{f}{{}^cz} = \frac{u}{{}^cx} = \frac{v}{{}^cy}.$$

Using homogeneous coordinates the projection law can be linearized and rewritten using matrix notation. The transformation matrix for projective

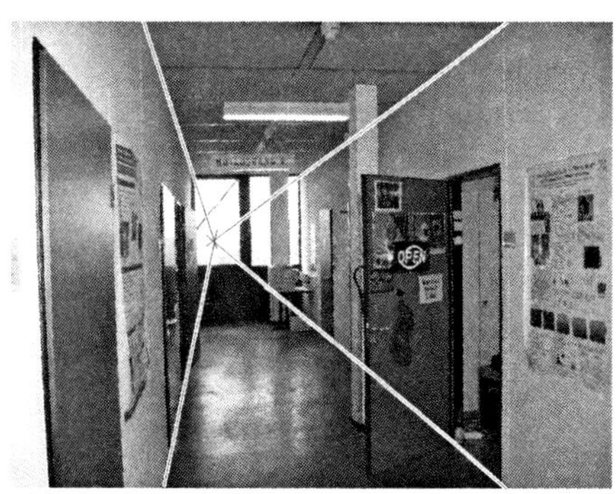

Fig. 2.9 As a consequence of the projective geometry parallel lines intersect in vanishing points

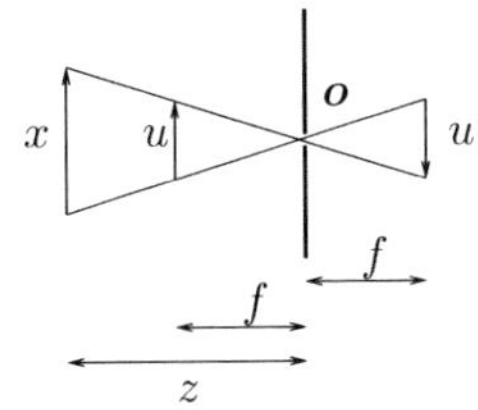

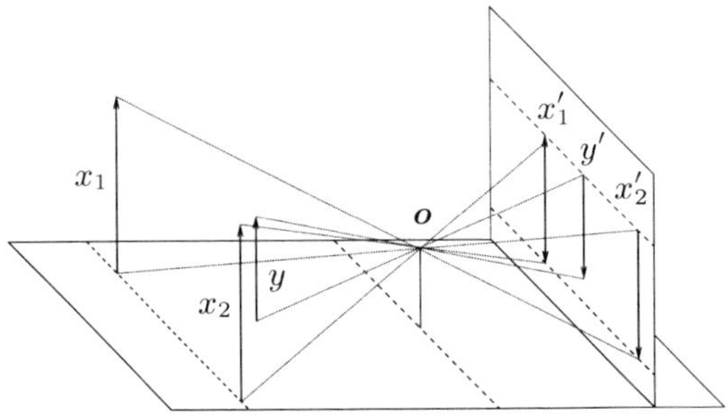

Fig. 2.10 The pinhole camera model

Fig. 2.11 The camera coordinate system is fixed to the camera. It is defined by the optical axis and image center.

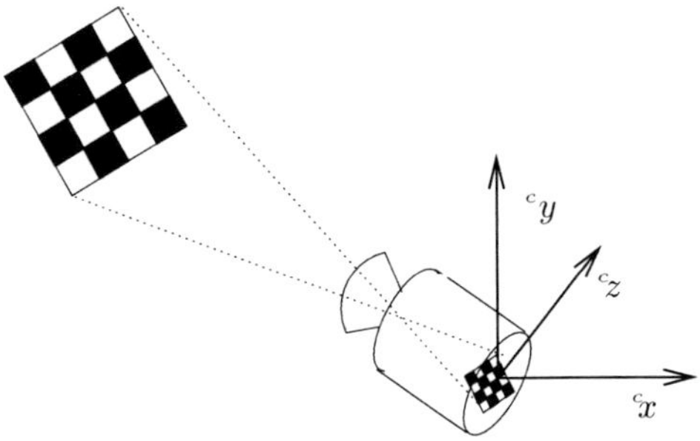

transformation from homogeneous 3D coordinates to homogeneous 2D coordinates is given as follows:

$$\begin{pmatrix} u' \\ v' \\ w' \end{pmatrix} = \begin{pmatrix} f & 0 & 0 & 0 \\ 0 & f & 0 & 0 \\ 0 & 0 & 1 & 0 \end{pmatrix} \begin{pmatrix} {}^c x \\ {}^c y \\ {}^c z \\ 1 \end{pmatrix}, \tag{2.2}$$

The image coordinates (u, v) are calculated afterwards by

$$u = {}^{u'}\!/_{w'}$$
$$v = {}^{v'}\!/_{w'}.$$

Usually the optical axis intersects the sensor at the coordinates u_0, v_0. In addition the sensor often exhibits a shear s and the focal length is determined in pixel units, i.e., $\alpha_x = f\ k_x$ and $\alpha_y = f\ k_y$ where the constants k_x and k_y describe the pixel sizes of the camera in x and y direction. Then Eq. (2.2) changes to

$$\begin{pmatrix} u' \\ v' \\ w' \end{pmatrix} = \begin{pmatrix} \alpha_x & s & u_0 & 0 \\ 0 & \alpha_y & v_0 & 0 \\ 0 & 0 & 1 & 0 \end{pmatrix} \begin{pmatrix} {}^c x \\ {}^c y \\ {}^c z \\ 1 \end{pmatrix}, \tag{2.3}$$

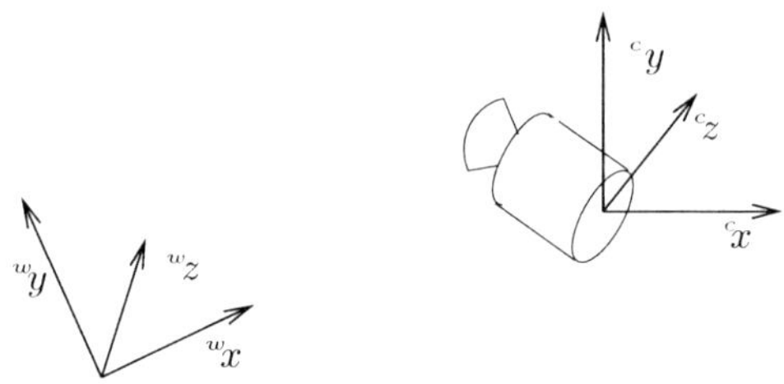

Fig. 2.12 The wolrd coordinate system describes objects in space and must be transformed into a camera coordinate system

The entries of the projection matrix in (2.3) are called intrinsic parameters and are determined by camera calibration. Besides the intrinsic parameters there are extrinsic camera parameters, that represent the pose (position and orientation) of the 3D camera coordinate system with respect to a 3D world coordinate system. The extrinsic parameters are given by a rotation matrix $\boldsymbol{R}$ and a translation vector $\boldsymbol{t}$. It holds:

$$\begin{pmatrix} {}^{c}x \\ {}^{c}y \\ {}^{c}z \\ 1 \end{pmatrix} = \begin{pmatrix} & \boldsymbol{R} & & \boldsymbol{t} \\ 0 & 0 & 0 & 1 \end{pmatrix} \begin{pmatrix} {}^{w}x \\ {}^{w}y \\ {}^{w}z \\ 1 \end{pmatrix} .$$

The pinhole camera model is also a model for imaging using a thin lens (see Figure 2.13) and the formulas given so far are applicable. This simplification does not regard aberrations that occur due to the use of lenses. The aberrations, i.e., deviations from the ideal pinhole model, can be divided into two categories. The first category consists geometric aberrations and include spherical aberration, astigmatism, distortion and coma. The second class contain the chromatic aberrations and deal with errors due to the refractive index function of the wavelength.

Geometric aberrations can be handled and reduced by a combination of lenses and by software. Figure 2.14 shows a typical image of a webcam and an undistorted version.

Camera calibration takes care of spherical aberration. Besides the intrinsic and extrinsic constants additional parameters that describe the radial and tangential distortion are calculated through calibration. The reason for radial

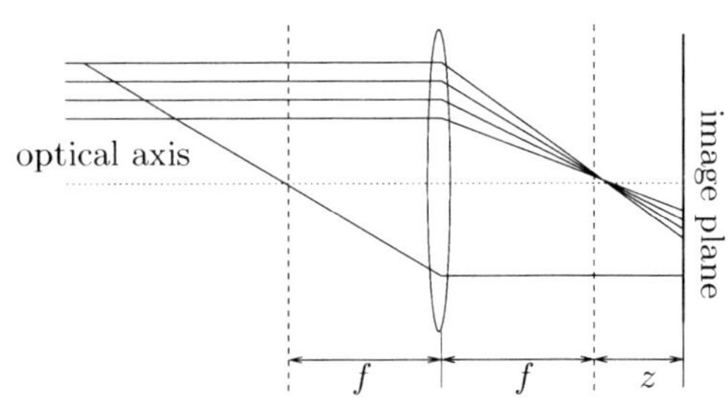

Fig. 2.13 The thin lens law

Fig. 2.14 Camera calibration allows to exclude radial and tangential abberarion. Left: image taken by a cheap webcam. Right: Undistorted image. For calibration, a pattern that allows simple detection of features is used.

distortions is that parallel rays do not converge towards a single focal point. The outer areas of an image are projected with a smaller focal length. The following calculation in conjunction with interpolation solves the problem:

$$\begin{pmatrix} {}^{\mathrm{undist}}u \\ {}^{\mathrm{undist}}v \end{pmatrix} = \begin{pmatrix} u \\ v \end{pmatrix} \left(1 + d_1\, r^2 + d_2\, r^4\right). \tag{2.4}$$

Here r denotes the distance of the point to the image center, i.e., $r^2 = (u - u_0)^2 + (v - v_0)^2$. Tangential distortion shows a dependence of the distance to the image center, too, but displaces the point along a tangential line.

2.3.2 Stereo Cameras

Distance information can be computed using two cameras that are mounted in a fixed distance. Figure 2.15 (left) shows such a stereo camera setup. Fusing two images that are taken with two cameras enables one to compute

Fig. 2.15 Left: Stereo camera system with two digital cameras. Right: Disparity image that records the displacement of corresponding pixels as grey values. The disparities are inversely correlated with distances.

distances from the difference of corresponding image points. Disparity refers to the difference in image location of an object seen by the left and right camera. By triangulation based on known disparity $d = x_l - x_r$ and baseline b, i.e., distances of the two cameras the depth z of an image point is calculated (see Figure 2.16):

$$\frac{{}^{c}z}{{}^{c}z + f} = \frac{b}{b + x_l - x_r}$$

resp.

$${}^{c}z = \frac{b\,f}{x_l - x_r} = \frac{b\,f}{d}.$$

After the depth is computed, the coordinates ${}^{c}x$ and ${}^{c}y$ are determined as follows:

$$\frac{{}^{c}x + {}^{b}\!/_{2}}{{}^{c}z} = \frac{x_l}{f} \qquad \text{resp.} \qquad {}^{c}x = \frac{b}{2}\left(\frac{x_l + x_r}{d}\right)$$

and

$$\frac{{}^{c}y}{{}^{c}z} = \frac{y}{f} \qquad \text{resp.} \qquad {}^{c}y = \frac{b}{2}\left(\frac{2y}{d}\right).$$

The depth resolution of a stereo camera is the smallest distinguishable depth. It depends on the smallest distinguishable disparity d and the distance between the cameras b and the focal length f. Higher depth resolutions can be achieved by increasing the baseline b, using cameras with larger focal length f or increasing the number of pixels, i.e., decrease the size of the pixel. Besides the depth resolution, an important issue while working with stereo cameras are the blind areas, i.e., objects that are only visible from one camera or that are between the two cameras and close to the stereo rig cannot be seen. To reduce the blind area it is often necessary to place the cameras close to each other, which in turn reduces depth resolution.

In real world setups, canonical stereo geometry as presented in Figure 2.16 does not occur, since it is not possible to adjust the cameras such that the image sensors lie in one plane. The cameras will always be rotated and translated, which has to be determined by calibration of the stereo system. Afterwards, the images can be rectified, i.e., they are projected into one plane (cf. Figure 2.17).

The following two problems must be solved while processing stereo images:

Correspondence problem. The prerequisite for computing the disparity is the finding of corresponding pixel. For solving this problem, one can notice, that the two camera centers $(\boldsymbol{o}_l, \boldsymbol{o}_r)$ and the processed point $\boldsymbol{p}$ form a triangle, that is called epipolar plane. (see Figure 2.17). The projection of a camera center on the other camera plane is called epipole $(\boldsymbol{o}_l, \boldsymbol{o}_r)$. If

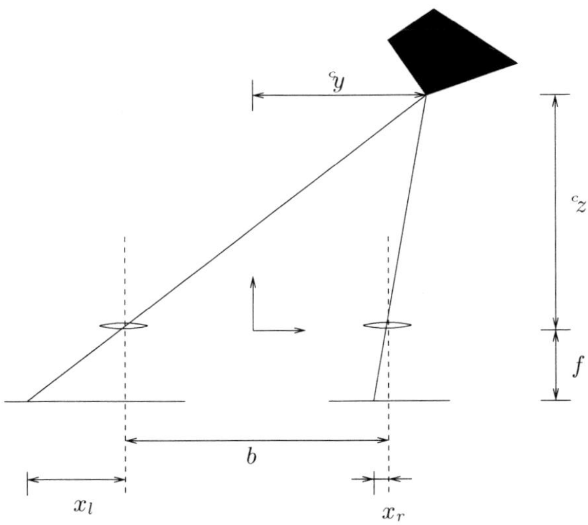

Fig. 2.16 Canonical stereo geometry

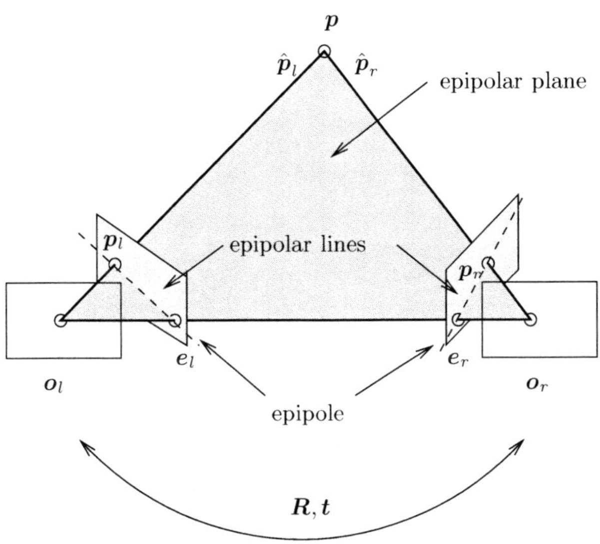

Fig. 2.17 General stereo geometry. Epipolar plane: plane that is defined by $\boldsymbol{o}$ and the two projection centers $\boldsymbol{o}_l, \boldsymbol{o}_r$. Epipol: Image of the projection center of one camera on the other camera plane. Epipolar constraint: Correspondence must be on the epipolar line.

the epipoles are known the search for corresponding pixel can be reduced to the search along the epipolar line.

Camera calibration. During camera calibration the extrinsic parameters of both cameras are determined. This gives the rotation and translation

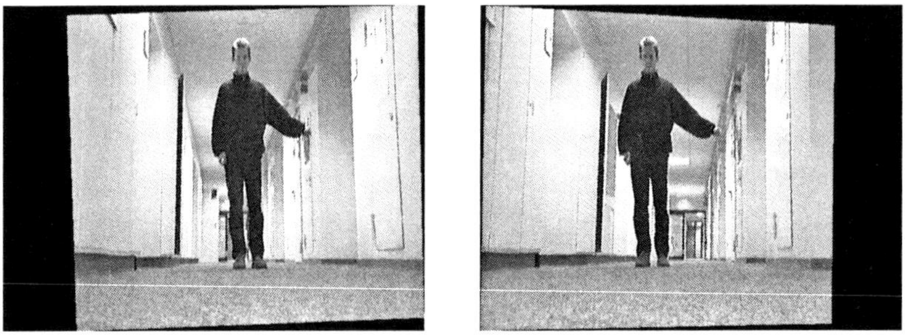

Fig. 2.18 Rectified stereo images allow to search for corresponding points along horizontal lines. The resulting disparity image is given in Figure 2.15 right.

$(\boldsymbol{R}, \boldsymbol{t})$ between the two cameras. These parameters allow the rectification of the images and simplify the correspondence problem: After rectification corresponding points must lie on the scan lines of the images. Therefore matching can be performed line-by-line.

Figure 2.18 shows a rectified stereo image pair. The next subsection describes the computation of the depth in detail.

Stereo Image Processing

To start lets assume that we have two cameras as given in Figure 2.17. For two corresponding points $\hat{\boldsymbol{p}}_l, \hat{\boldsymbol{p}}_r \in \mathbb{R}^3$ the 3D coordinates $\boldsymbol{p}_l, \boldsymbol{p}_r \in \mathbb{R}^3$ of the image points are known. A transformation of the camera coordinate systems yields the following equation:

$$\hat{\boldsymbol{p}}_r = \boldsymbol{R}(\hat{\boldsymbol{p}}_l - \boldsymbol{t}), \tag{2.5}$$

where the transformation consists of a rotation $\boldsymbol{R}$ and a translation $\boldsymbol{t}$. The camera centers $\boldsymbol{o}_l, \boldsymbol{o}_r$, the point $\hat{\boldsymbol{p}}$ as well as the projections of the centers lie in one plane. Therefore, the vectors $\boldsymbol{t}, \hat{\boldsymbol{p}}_l, \hat{\boldsymbol{p}}_l - \boldsymbol{t}$ are co-planar and it holds

$$0 = (\hat{\boldsymbol{p}}_l - \boldsymbol{t})^T(\boldsymbol{t} \times \hat{\boldsymbol{p}}_l). \tag{2.6}$$

Putting (2.5) in (2.6) yields

$$\begin{aligned} 0 &= (\boldsymbol{R}^T\hat{\boldsymbol{p}}_r)^T(\boldsymbol{t} \times \hat{\boldsymbol{p}}_l) \qquad \text{resp.} \\ &= (\boldsymbol{R}^T\hat{\boldsymbol{p}}_r)^T\boldsymbol{S}\hat{\boldsymbol{p}}_l \\ &= \hat{\boldsymbol{p}}_r^T\boldsymbol{R}\boldsymbol{S}\hat{\boldsymbol{p}}_l \end{aligned}$$

Here the cross-product is written as matrix-vector multiplication:

$$\boldsymbol{t} \times \hat{\boldsymbol{p}}_l = \boldsymbol{S}\hat{\boldsymbol{p}}_l \qquad \text{where} \quad \boldsymbol{T} = \begin{pmatrix} 0 & -t_z & t_y \\ t_z & 0 & -t_x \\ -t_y & t_x & 0 \end{pmatrix}$$

The multiplication $\boldsymbol{E} = \boldsymbol{T}\boldsymbol{R}$ is called essential matrix. It contains the wanted rotation and translation. For the image points $\boldsymbol{p}_l$ and $\boldsymbol{p}_r$ we reason in an analogous fashion:

$$\hat{\boldsymbol{p}}_r^T \boldsymbol{E} \hat{\boldsymbol{p}}_l = 0 \quad \Rightarrow \quad \boldsymbol{p}_r^T \boldsymbol{E} \boldsymbol{p}_l = 0$$

The essential matrix $\boldsymbol{E}$ features some properties, that are given by the following lemmas.

Lemma 1. *$\boldsymbol{t}_0$ is the eigenvector of $\boldsymbol{E}^T$, that corresponds to the 0 eigenvalue.*

Proof. It is:

$$\boldsymbol{E}^T \boldsymbol{t}_0 = (\boldsymbol{T}\boldsymbol{R})^T \boldsymbol{t}_0 = -\boldsymbol{R}^T \boldsymbol{T} \times \boldsymbol{t}_0 = -\boldsymbol{R}^T \boldsymbol{t}_0 \times \boldsymbol{t}_0 = \boldsymbol{R}^T \boldsymbol{0} = 0 \cdot \boldsymbol{T} \qquad \square$$

Lemma 2. *The singular value decomposition of $\boldsymbol{E}$ yields $\boldsymbol{E} = \boldsymbol{U}\boldsymbol{S}\boldsymbol{V}$ with the singular values $\sigma_0 = 1$, $\sigma_1 = 1$ and $\sigma_2 = 0$.*

Proof. See [64]. □

The singular value decomposition $\boldsymbol{E} = \boldsymbol{U}\boldsymbol{S}\boldsymbol{V}$ yields orthonormal matrices $\boldsymbol{U}$ and $\boldsymbol{V}$. $\boldsymbol{S}$ is a diagonal matrix that contains the singular values σ_i:

$$\boldsymbol{S} = \begin{pmatrix} \sigma_0 & 0 & 0 \\ 0 & \sigma_1 & 0 \\ 0 & 0 & \sigma_3 \end{pmatrix}$$

Lemma 3. *It holds : $\|\boldsymbol{E}\|^2 = tr\left(\boldsymbol{E}^T \boldsymbol{E}\right) = 2$*

Proof. See [64]. □

Unfortunately the 3D coordinates $\boldsymbol{p}_l, \boldsymbol{p}_r \in \mathbb{R}^3$ are usually unknown. Only the 2D image coordinates $\bar{\boldsymbol{p}}_l = (u_l, v_l, 1)^T \in \mathbb{R}^3$ and $\bar{\boldsymbol{p}}_r = (u_r, v_r, 1)^T \in \mathbb{R}^3$ are given in the image coordinate systems. The transition from the essential matrix $\boldsymbol{E}$ to the fundamental matrix $\boldsymbol{F}$ regards this issue.

$$\boldsymbol{p}_r^T \boldsymbol{E} \boldsymbol{p}_l = 0 \quad \Rightarrow \quad \bar{\boldsymbol{p}}_r^T \boldsymbol{F} \bar{\boldsymbol{p}}_l = 0.$$

The fundamental matrix models the mapping at the camera (see. Eq. (2.3)):

$$\boldsymbol{F} = \boldsymbol{M}_r^{-1} \boldsymbol{E} \boldsymbol{M}_l^{-1} \quad \text{where} \quad \boldsymbol{M} = \begin{pmatrix} \alpha_x & 0 & u_0 \\ 0 & \alpha_y & v_0 \\ 0 & 0 & 1 \end{pmatrix}.$$

The most common algorithm for computing the matrix $\boldsymbol{F}$ is the 8 point algorithm. Input are $n \geq 8$ point correspondences that lead to the minimization of

$$\operatorname{argmin}_{\boldsymbol{F}} (\bar{\boldsymbol{p}}_r \boldsymbol{F} \bar{\boldsymbol{p}}_l)^2$$

The algorithm aims at finding the matrix $\boldsymbol{F}$ that minimizes the point distances. This problem is reduced to a least square problem and the following term is minimized:

$$\min_{\boldsymbol{F}} \|\boldsymbol{A}\boldsymbol{F}\|$$

The matrix $\boldsymbol{A}$ above is defined as

$$\boldsymbol{A} = \begin{pmatrix} u_{l,1}u_{r,1} & u_{l,1}v_{r,1} & u_{l,1} & v_{l,1}u_{r,1} & v_{l,1}v_{r,1} & v_{l,1} & u_{r,1} & v_{r,1} & 1 \\ u_{l,2}u_{r,2} & u_{l,2}v_{r,2} & u_{l,2} & v_{l,2}u_{r,2} & v_{l,2}v_{r,2} & v_{l,2} & u_{r,2} & v_{r,2} & 1 \\ & & & \vdots & & & & & \\ u_{l,n}u_{r,n} & u_{l,n}v_{r,n} & u_{l,n} & v_{l,n}u_{r,n} & v_{l,n}v_{r,n} & v_{l,n} & u_{r,n} & v_{r,n} & 1 \end{pmatrix}$$

Every point pair yields an equation. Singular value decomposition is used to solve the minimization problem: $\boldsymbol{A} = \boldsymbol{U}\boldsymbol{S}\boldsymbol{V}$, where $\boldsymbol{U}$ and $\boldsymbol{V}$ are orthonormal matrices. $\boldsymbol{S}$ is a diagonal matrix storing the singular values σ_i:

$$\boldsymbol{S} = \begin{pmatrix} \sigma_0 & 0 & \cdots & 0 \\ 0 & \sigma_1 & \cdots & 0 \\ & \vdots & \ddots & 0 \\ 0 & 0 & \cdots & \sigma_9 \end{pmatrix}$$

The singular values are sorted, i.e., $\sigma_0 > \sigma_1 > \cdots > \sigma_9$. Using exactly 8 points yields $\sigma_9 = 0.$, since 8 equations and 9 unknown variables are present. The computed matrix $\boldsymbol{F}_{\text{est}} = \boldsymbol{V}$ has to be revised, since we have to ensure rank 2 (see Lemma 2). SVD is applied again, this time to the matrix $\boldsymbol{F}_{\text{est}}$:

$$\boldsymbol{F}_{\text{est}} = \boldsymbol{U}\boldsymbol{S}\boldsymbol{V} = \boldsymbol{U} \begin{pmatrix} \sigma_0 & 0 & 0 \\ 0 & \sigma_1 & 0 \\ 0 & 0 & \sigma_2 \end{pmatrix} \boldsymbol{V}$$

The smallest singular value is set to zero ($\sigma_2 = 0$) and the resulting matrix is called $\boldsymbol{S}'$. Finally the fundamental matrix is given by

$$\boldsymbol{F} = \boldsymbol{U}\boldsymbol{S}'\boldsymbol{V}.$$

After computing the fundamental matrix $\boldsymbol{F}$ images can be rectified. Figure 2.18 and 2.19 show rectified image pairs, where correspondences lie on horizontal image scan lines. Unfortunately determining unique correspondences

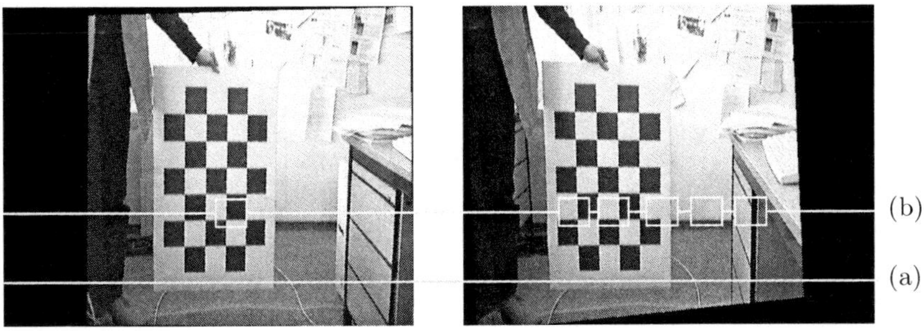

Fig. 2.19 Rectified stereo images allow searching for corresponding pixels along a horizontal line. Since mans pixel values along the line (a) are identical, one compares regions (b).

is still hard. For every pixel in one image there are several, having a similar or same grey value. In addition the same object might be perceived with different illuminations, since the cameras capture different views. Therefore, stereo algorithms do no not match pixel, but compare image regions instead (cf. Figure 2.19). To compare image regions, different correlation methods have been developed, e.g.,

$$\sum_{[i,j]\in R} (f(i,j) - g(i,j))^2 \tag{2.7}$$

$$\sum_{[i,j]\in R} f(i,j)g(i,j). \tag{2.8}$$

Here the images are f and g. The term (2.7) is called sum of squared distances (SSD) and has to be minimized. If image regions are identical the numerical value is 0. The term (2.8) is the cross correlation (CC) and shows a maximum for identical image regions.

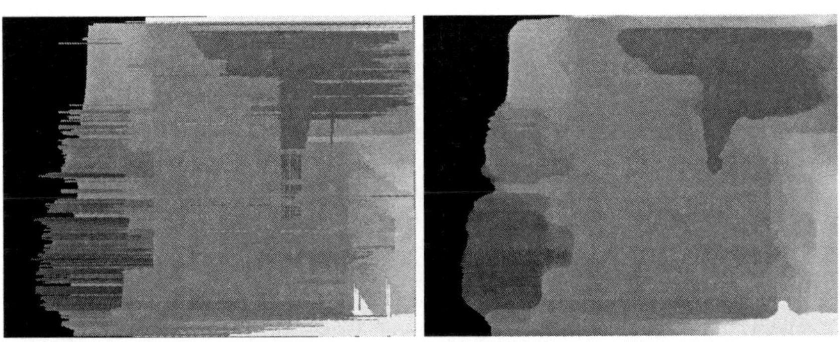

Fig. 2.20 Disparity images based on the rectified images of Figure 2.19: Left: Results using a small search window. Right: large search window.

Figure 2.20 shows two disparity maps. Comparing image regions yields an approximation, since image regions might be visible only in one camera and might not have correspondences. Here the influence of the size of the image region is visible.

Besides comparing regions there are feature based approaches that aim at finding corresponding features, for instance SIFT features. The depth is calculated only at the feature points and sparse range maps are created.

2.4 3D Cameras

A current trend is the development of 3D cameras (see Figure 2.21). These cameras contain a CMOS sensor and additional laser diodes that illuminate the scene with modulated light. The CMOS sensor detects the light and computes on die the time-of-flight based on the detected phase shift. The output of this sensor are range values and reflectance values, i.e., the amount of light returned to the sensors (cf. Figure 2.22).

Similarly to a normal camera a 3D camera can be calibrated. Calibration determines the intrinsic parameters (the focal length (α_x and α_y), the image center (u_0, v_0) and the distortion coefficients (d_1, d_2). Afterwards the 3D coordinates are calculated directly: First of all the distortion is removed (see Eq. (2.4)), i.e., $({}^{\text{undist}}u, {}^{\text{undist}}v)$ are computed. Then the 3D coordinates in the camera coordinate system $({}^{c}x, {}^{c}y, {}^{c}z)$ are given by solving

$$
\begin{aligned}
{}^{\text{undist}}u &= \alpha_x \frac{{}^{c}x}{{}^{c}z} + c_x \\
{}^{\text{undist}}v &= \alpha_y \frac{{}^{c}y}{{}^{c}z} + c_y \\
r &= \sqrt{{}^{c}x^2 + {}^{c}y^2 + {}^{c}z^2}
\end{aligned}
$$

based on the distances r. This process yields a 3D point cloud.

Fig. 2.21 3D Cameras: Left Swissranger manufactured by MESA Imaging AG. It uses technology developed by CSEM. Right: PMD (Photonic Mixer Device) Camera by PMDTechnologies GmbH. Figures courtesy of MESA Imaging and PMDTechnologies.

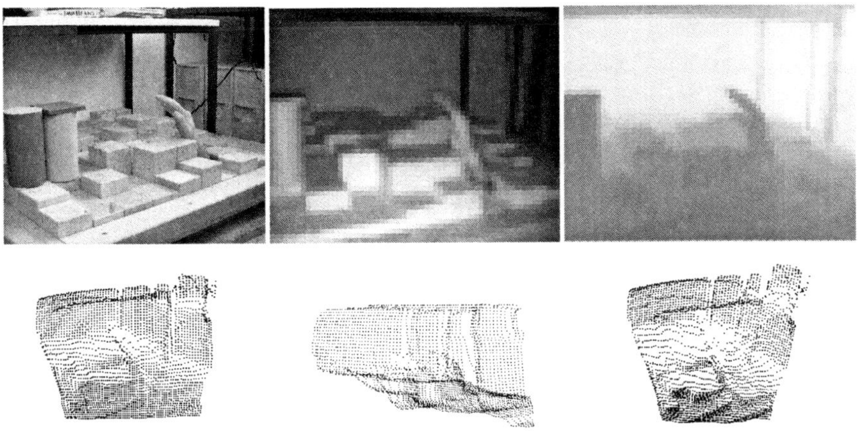

Fig. 2.22 Captured test scene. Top row: Left: Image taken with a digital camera. Middle: Reflectance image acquired by the O3D100. Right: Depth image acquired by the O3D100. Bottom row: 3D point cloud from different perspectives.

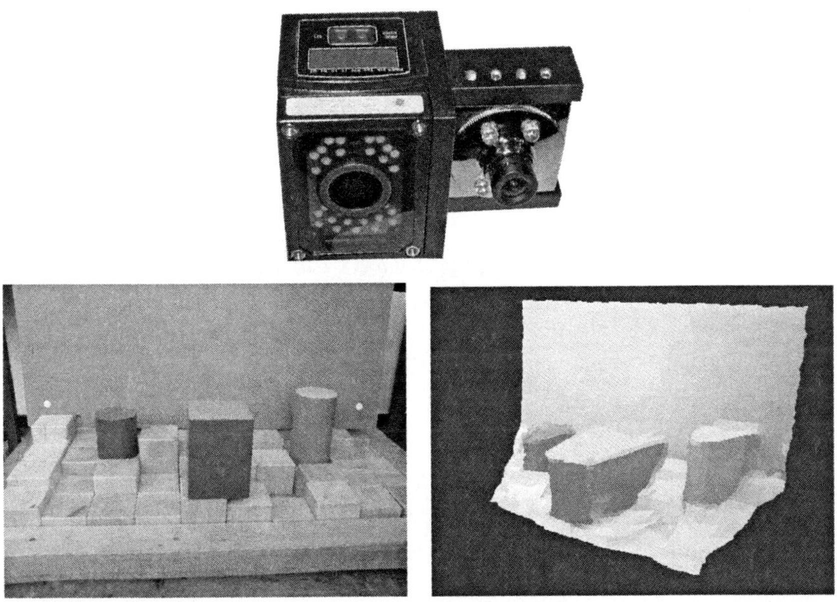

Fig. 2.23 Top: IFM O3D100 in combination with a webcam. Bottom: Test scene and textured 3D image.

A combination of a normal camera and a 3D camera utilizes the advantages of both. Figure 2.23 presents 3D camera (IFM O3D100) in combination with a webcam and typical images.

Chapter 3
State of the Art

3.1 Categorization of Robotic Mapping Algorithms

One way to categorize mapping algorithms is by the map type. The map can be topological or metrical. Metrical maps represent quantitive distances of the environment. These maps are either 2D, usually in an upright projection, or 3D, i.e., a volumetric environment map. Furthermore, SLAM approaches can be classified by the number of DoF of the robot pose. A 3D pose estimate contains the (x, y)-coordinate and a rotation θ, whereas a 6D pose estimate considers all DoF a rigid mobile robot can have, i.e., the (x, y, z)-coordinate and the roll, yaw and pitch angles.

In the literature, three different techniques are used for generating 3D maps: First, a planar localization method combined with a 3D sensor; second, a precise 6D pose estimate combined with the data from a 2D sensor; and third, a 3D sensor with a 6D localization method. Table 3.1 summarizes these mapping techniques (black) in comparison with planar 2D mapping (grey). In this book, we focus on 3D data and 6D localization, hence on 6D SLAM.

Figure 3.1 shows an example of a robot capable of 2D mapping with the corresponding 2D map. In this case, a grid map has been used, where every grid cell contains the probability that the cell is occupied. Figure 3.2 presents a state of the art robot for planar 3D mapping and corresponding maps. Figure 3.3 presents the robot Stanley. Several fixed but tilted mounted laser scanner (SICK LMS 291) acquire slices of the environment, and therefore we speak about Slice-wise 6D SLAM.

3.1.1 Planar 2D Mapping

State of the art for planar 2D metric maps are probabilistic methods, where the robot has probabilistic motion and uncertain perception models. By integrating these two distributions with a Bayes filter, e.g., Kalman or particle filter, it is possible to localize the robot. Mapping is often an extension to this

A. Nüchter: 3D Robotic Map.: The Simultaneous Local. & Map. Prob., STAR 52, pp. 29–33.
springerlink.com

Table 3.1 Overview of the dimensionality of SLAM approaches. Grey: 2D maps. Black: 3D maps.

		Dimensionality of pose representation 3D	Dimensionality of pose representation 6D
Sensor data	2D	**Planar 2D mapping** 2D mapping of planar sonar and laser scans. See [120] for a detailed overview.	**Slice-wise 6D SLAM** 3D mapping using a prec ise localization, considering the x,y,z-position and the roll yaw and pitch angle.
Sensor data	3D	**Planar 3D mapping** 3D mapping using a planar localization method and, e.g., an upward looking laser scanner or 3D scanner.	**Full 6D SLAM** 3D mapping using 3D laser sca nners or (stereo) cameras with pose estimates calculated from the sensor data.

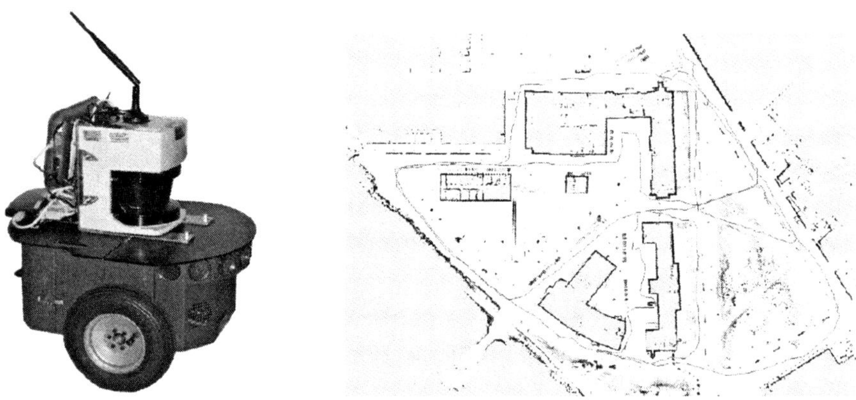

Fig. 3.1 Left: Pioneer DX-2 robot equipped with a SICK laser scanners (LMS 291) for 2D mapping. Right: Acquired 2D map. Figure courtesy of G. Grisetti et al. [51].

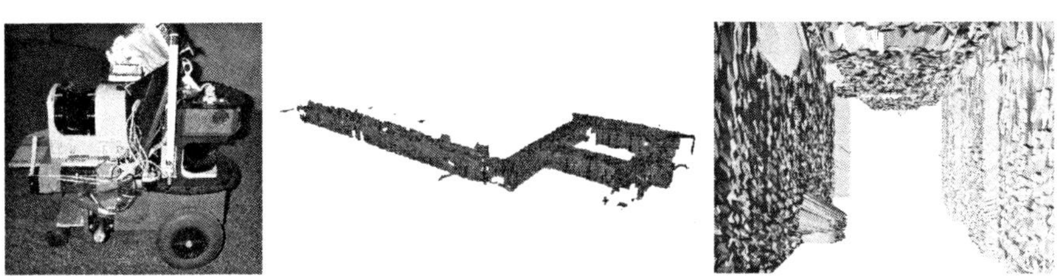

Fig. 3.2 Left: Pioneer DX-2 robot equipped with two SICK laser scanner (PLS 100) for 3D mapping. The upward pointing scanner is used to acquire 3D data. Middle and right: Acquired 3D maps. Figure courtesy of S. Thrun.

Fig. 3.3 Slice-wise 6D SLAM on the robot Stanley. Stanley won the DARPA Grand Chellange 2005. Figure courtesy of S. Thrun.

estimation problem. Besides the robot pose, positions of landmarks are estimated. Closed loops, i.e., a second encounter of a previously visited area of the environment, play a special role here. Once detected, they enable the algorithms to bound the error by deforming the already mapped area such that a topologically consistent model is created. However, there is no guarantee for a correct model. Several strategies exist for solving SLAM. Thrun reviews in [120] existing techniques, i.e., maximum likelihood estimation [38], expectation maximization [34, 121], extended Kalman filter [26] or (sparse extended) information filter [124]. In addition to these methods, FastSLAM [122], which approximates the posterior probabilities, i.e., robot poses, by particles, and the method of Lu/Milios on the basis of IDC scan matching [78], play an important role in 2D.

3.1.2 Planar 3D Mapping

Instead of using 3D scanners, which yield consistent 3D scans in the first place, some groups have attempted to build 3D volumetric representations of environments by translating 2D laser range finders. Thrun et al. [122], Früh et al. [42] and Zhao et al. [141] use two 2D laser scanners for acquiring 3D data. One scanner is mounted horizontally, another vertically. The latter one grabs a vertical scan line which is transformed into 3D points based on the current 3D robot pose. Since the vertical scanner is not able to scan sides of objects, Zhao et al. use two additional, vertically mounted 2D scanners, shifted by 45° to reduce occlusions [141]. The horizontal scanner is used to compute the 3D robot pose. The precision of 3D data points depends, besides on the precision of the scanner, critically on that pose estimation.

Recently, different groups employ rotating SICK scanners for acquiring 3D data [69, 117, 136]. Wulf et al. let the scanner rotate around the vertical axis. They acquire 3D data while moving, thus the quality of the resulting map crucially depends on the pose estimate that is given by inertial sensors, i.e., gyros [136]. In addition, their SLAM algorithms do not consider all 6 DoF.

3.1.3 Slice-Wise 6D SLAM

Local 3D maps built by translated 2D laser scanners and 6D pose estimates are often used for mobile robot navigation. A well-known example is the grand challenge, where the Stanford racing team used this technique for high speed terrain classification [125].

3.1.4 Full 6D SLAM

A few other groups use highly accurate, yet somewhat immobile 3D laser scanners [2, 46, 111]. The RESOLV project aimed at modeling interiors for virtual reality and tele-presence [111]. They used a RIEGL laser range finder on robots and the ICP algorithm for scan matching [11]. The AVENUE project develops a robot for modeling urban environments [2], using a CYRAX scanner and a feature-based scan matching approach for registering the 3D scans. However, in their recent work they do not use data of the laser scanner in the robot control architecture for localization [46]. The group of M. Hebert has reconstructed environments using the Zoller+Fröhlich laser scanner and aims to build 3D models without initial position estimates, i.e., without odometry information [57]. Magnusson and Duckett proposed a 3D scan alignment method that – in contrast to the previously mentioned research groups – does not use ICP algorithm, but the normal distribution transform instead [79].

In principle, the probabilistic methods from planar 2D mapping are extendable to 3D mapping with 6D pose estimates. However, no reliable feature extraction nor a strategy for reducing the computational costs of multi hypothesis tracking, e.g., FastSLAM, that grows exponentially with the DoF, has been published to our knowledge.

3.1.5 Recent Trends

A recent trend in SLAM research is to apply probabilistic methods to 3D mapping. Katz et al. use a probabilistic notion of ICP scan matching [67]. Weingarten et al. [133] and Cole et al. [19] use an extended Kalman filter on the mapping problem. In the present paper, we extend this state of the art by a GraphSLAM method. A similar approach was used in [127]. However, their algorithm is not practical due to the reported computational requirements. Without considering any sensor data such as laser scans, Olson et al. create globally consistent maps by minimizing the global non-linear constraint network on a set of poses [93]. Triebel et al. apply this approach to their multi-level surface maps [128]. Furthermore, Frese presented an extension of his treemap SLAM algorithm to 6 DoF [37].

3.2 Globally Consistent Range Image Alignment

Besides the robotic community, computer vision researchers are interested in consistent alignment methods. Chen and Medoni [18] aimed at globally consistent range image alignment when introducing an incremental matching method, i.e., all new scans are registered against the so-called metascan, which is the union of the previously acquired and registered scans. This method does not spread out the error and is order-dependent.

Bergevin et al. [10], Benjemaa and Schmitt [7, 8], and Pulli [102] present iterative approaches. Based on networks representing overlapping parts of images, they use the ICP algorithm for computing transformations that are applied after all correspondences between all views have been found. However, the focus of research is mainly 3D modeling of small objects using a stationary 3D scanner and a turn table; therefore, the used networks consist mainly of one loop [102]. A probabilistic approach was proposed by Williams et al.[134], where each scan point is assigned a Gaussian distribution in order to model the statistical errors made by laser scanners. This requires long computation time due to the large amount of data in practice. Krishnan et al. [70] presented a global registration algorithm that minimizes the global error function by optimization on the manifold of 3D rotation matrices.

Chapter 4
3D Range Image Registration

Multiple 3D scans are necessary to digitalize environments without occlusions. To create a correct and consistent model, the scans have to be merged into one coordinate system. This process is called registration. If the localization of the robot with the 3D scanner were precise, the registration could be done directly by the robot pose. However, due to unprecise robot sensors, the self localization is erroneous, so the geometric structure of overlapping 3D scans has to be considered for registration.

Scan matching approaches can be classified into two categories:

Matching as an optimization problem uses a cost function for the quality of the alignment of the scans. The range images are registered by determining the rigid transformation (rotation and translation) which minimizes the cost function.

Feature based matching extracts distinguishing features of the range images and uses corresponding features for calculating the alignment of the scans.

4.1 The ICP Algorithm

The following method is used for registration of point sets. The complete algorithm was invented at the same time in 1991 by Besl and McKay [11], by Chen and Medioni [17] and by Zhang [140]. The method is called *Iterative Closest Points (ICP) algorithm.*

Given two independently acquired sets of 3D points, $\hat{M}$ (model set, $|\hat{M}| = N_m$) and $\hat{D}$ (data set, $|\hat{D}| = N_d$) which correspond to a single shape, we want to find the transformation $(\boldsymbol{R}, \boldsymbol{t})$ consisting of a rotation matrix $\boldsymbol{R}$ and a translation vector $\boldsymbol{t}$ which minimizes the following cost function:

A. Nüchter: 3D Robotic Map.: The Simultaneous Local. & Map. Prob., STAR 52, pp. 35–75.
springerlink.com

$$E(\boldsymbol{R},\boldsymbol{t}) = \sum_{i=1}^{N_m}\sum_{j=1}^{N_d} w_{i,j} \left|\left|\hat{\boldsymbol{m}}_i - (\boldsymbol{R}\hat{\boldsymbol{d}}_j + \boldsymbol{t})\right|\right|^2 . \tag{4.1}$$

$w_{i,j}$ is assigned 1 if the i-th point of $\hat{M}$ describes the same point in space as the j-th point of $\hat{D}$. Otherwise $w_{i,j}$ is 0. Two things have to be calculated: First, the corresponding points, and second, the transformation ($\boldsymbol{R}$, $\boldsymbol{t}$) that minimizes $E(\boldsymbol{R},\boldsymbol{t})$ on the base of the corresponding points.

The ICP algorithm calculates iteratively the point correspondences. In each iteration step, the algorithm selects the closest points as correspondences and calculates the transformation $(\boldsymbol{R},\boldsymbol{t})$ for minimizing equation (4.1). Besl et al. proves that the method terminates in a minimum [11]. The assumption is that in the last iteration step the point correspondences are correct.

Algorithm 4.1. The ICP algorithm.

1: **for** $i = 0$ to *maxIterations* **do**
2: **for all** $\boldsymbol{d}_j \in D$ **do**
3: find the closest point within a range $d_{\max}$ in the set M for point $\boldsymbol{d}_j$
4: **end for**
5: Calculate transformation $(\boldsymbol{R},\boldsymbol{t})$ that minimizes the error function Eq. (4.1)
6: Apply the transformation found in step 5 to the data set D.
7: Compute the difference of the quadratic error, i.e., compute the difference of the value $||E_{i-1}(\boldsymbol{R},\boldsymbol{t}) - E_i(\boldsymbol{R},\boldsymbol{t})||$ before and after the application of the transformation. If this difference falls below a threshold ε, terminate.
8: **end for**

Besl et al. prove that the method terminates in a minimum [11, 33]: All steps of the ICP algorithm reduce the error: $E_{i+1}(\boldsymbol{R},\boldsymbol{t}) < E_i(\boldsymbol{R},\boldsymbol{t})$. By the monotone convergence theorem ICP terminates in a minimum, since the error is bounded. However, this theorem does not hold in our case, since we use a maximum tolerable distance $d_{\max}$ for associating the scan data. Such a threshold is required, given that the 3D scans overlap only partially. After a transformation some data points that did not have a closest point within this limit might have one and therefore the value of the error function might increase. Finally, the algorithm minimizes Eq. (4.1) *and* maximizes the number of corresponding points.

4.1.1 Direct Solutions of the ICP Error Function

In every iteration the optimal transformation ($\boldsymbol{R}$, $\boldsymbol{t}$) for minimizing the error function $E(\boldsymbol{R},\boldsymbol{t})$ in Eq. (4.1) has to be computed. There exist different strategies to find the minimum that can be divided into direct and indirect methods.

The simulation of a physical spring system belongs to the indirect methods [27, 116]. Here the distances between the closest points that are summed in Eq. (4.1) are treated as springs that move the second point set into a pose with minimal energy. Similarly computations based on gradient descent, the Gausss-Newton method and the Levenberg-Marquardt algorithms belong to the category of indirect methods. Indirect methods have the disadvantage that they need to perform several evaluations of the function (4.1). Therefore, they need inevitably more time than direct methods. direct methods are also called closed form solutions.

Simplification of the Error Function

The double summation in Eq. (4.1) can be rewritten as

$$E(\boldsymbol{R},\boldsymbol{t}) = \sum_{i=1}^{N_m}\sum_{j=1}^{N_d} w_{i,j} \left|\left|\hat{\boldsymbol{m}}_i - (\boldsymbol{R}\hat{\boldsymbol{d}}_j + \boldsymbol{t})\right|\right|^2$$
$$= \frac{1}{N}\sum_{i=1}^{N} \left|\left|\boldsymbol{m}_i - (\boldsymbol{R}\boldsymbol{d}_i + \boldsymbol{t})\right|\right|^2, \qquad \text{using} \quad N = \sum_{i=1}^{N_m}\sum_{j=1}^{N_d} \operatorname{sgn} w_{i,j} \tag{4.2}$$

All corresponding points can be represented in a tuple $(\boldsymbol{m}_i, \boldsymbol{d}_i)$.

Note 1. Implementations of the ICP algorithm use Eq. (4.2). Instead of storing a matrix $\boldsymbol{W}$ for the weights a vector of point correspondences is built. There is no influence to following computations; the amount of memory is usually reduced from $\mathcal{O}(n^2)$ to $\mathcal{O}(n)$.

Closed Form Solution

Four algorithms are currently known that solve the error function of the ICP algorithm in closed form [76]. The difficulty of this minimization is to enforce the orthonormality constraint for the rotation matrix $\boldsymbol{R}$. In the following the four algorithms are are presented. The first three algorithms separate the computation of the rotation $\boldsymbol{R}$ from the computation of the translation $\boldsymbol{t}$. These algorithms compute the rotation first and afterwards the translation is derived using the rotation. For this a separation, two point sets M' und D' have to be computed, by subtracting the mean of the points that are used in the matching:

$$\boldsymbol{c}_m = \frac{1}{N}\sum_{i=1}^{N} \boldsymbol{m}_i, \qquad \boldsymbol{c}_d = \frac{1}{N}\sum_{i=1}^{N} \boldsymbol{d}_i \tag{4.3}$$

and

$$M' = \{\boldsymbol{m}'_i = \boldsymbol{m}_i - \boldsymbol{c}_m\}_{1,\ldots,N}, \qquad D' = \{\boldsymbol{d}'_i = \boldsymbol{d}_i - \boldsymbol{c}_d\}_{1,\ldots,N}. \tag{4.4}$$

After replacing Eq. (4.3) and (4.4) in the error function, $E(\boldsymbol{R}, \boldsymbol{t})$ Eq. (4.2) becomes:

$$E(\boldsymbol{R}, \boldsymbol{t}) = \frac{1}{N} \sum_{i=1}^{N} ||\boldsymbol{m}'_i - \boldsymbol{R}\boldsymbol{d}'_i - \underbrace{(\boldsymbol{t} - \boldsymbol{c}_m + \boldsymbol{R}\boldsymbol{c}_d)}_{=\tilde{\boldsymbol{t}}}||^2 \tag{4.5}$$

$$= \frac{1}{N} \sum_{i=1}^{N} ||\boldsymbol{m}'_i - \boldsymbol{R}\boldsymbol{d}'_i||^2 - \frac{2}{N} \tilde{\boldsymbol{t}} \cdot \sum_{i=1}^{N} (\boldsymbol{m}'_i - \boldsymbol{R}\boldsymbol{d}'_i) + \frac{1}{N} \sum_{i=1}^{N} ||\tilde{\boldsymbol{t}}||^2. \tag{4.6}$$

In order to minimize the sum above, all terms have to be minimized. The second sum is zero, since all values refer to centroid. The third part has its minimum for $\tilde{\boldsymbol{t}} = \boldsymbol{0}$ or

$$\boldsymbol{t} = \boldsymbol{c}_m - \boldsymbol{R}\boldsymbol{c}_d. \tag{4.7}$$

Therefore the algorithm has to minimize only the first term, and the error function is expressed in terms of the rotation only:

$$E(\boldsymbol{R}, \boldsymbol{t}) \propto \sum_{i=1}^{N} ||\boldsymbol{m}'_i - \boldsymbol{R}\boldsymbol{d}'_i||^2. \tag{4.8}$$

Computing the Transformation using the Singular Value Decomposition of a Matrix

The following method was developed 1987 by Arun, Huang und Blostein [3]. The rotation $\boldsymbol{R}$ is represented as an orthonormal 3×3 matrix.

Theorem 1. *The optimal rotation is calculated by* $\boldsymbol{R} = \boldsymbol{V}\boldsymbol{U}^T$. *Hereby the matrices* $\boldsymbol{V}$ *and* $\boldsymbol{U}$ *are derived by the singular value decomposition* $\boldsymbol{H} = \boldsymbol{U}\boldsymbol{\Lambda}\boldsymbol{V}^T$ *of a correlation matrix* $\boldsymbol{H}$. *This* 3×3 *matrix* $\boldsymbol{H}$ *is given by*

$$\boldsymbol{H} = \sum_{i=1}^{N} \boldsymbol{m}'^T_i \boldsymbol{d}'_i = \begin{pmatrix} S_{xx} & S_{xy} & S_{xz} \\ S_{yx} & S_{yy} & S_{yz} \\ S_{zx} & S_{zy} & S_{zz} \end{pmatrix}, \tag{4.9}$$

where $S_{xx} = \sum_{i=1}^{N} m'_{x,i} d'_{x,i}$, $S_{xy} = \sum_{i=1}^{N} m'_{x,i} d'_{y,i}$, ... *The analogous algorithm is derived directly from this theorem.*

Proof. Since a rotation is length preserving, i.e., $||\boldsymbol{R}\boldsymbol{d}'_i||^2 = ||\boldsymbol{d}'_i||^2$ the error function (4.8) is expanded

$$E(\boldsymbol{R},\boldsymbol{t}) = \sum_{i=1}^{N} ||\boldsymbol{m}_i'||^2 - 2\sum_{i=1}^{N} \boldsymbol{m}_i' \cdot \boldsymbol{R}\boldsymbol{d}_i' + \sum_{i=1}^{N} ||\boldsymbol{d}_i'||^2 .$$

The rotation affects only the middle term, thus it is sufficient to maximize

$$\sum_{i=1}^{N} \boldsymbol{m}_i' \cdot \boldsymbol{R}\boldsymbol{d}_i' = \sum_{i=1}^{N} \boldsymbol{m}_i'^T \boldsymbol{R}\boldsymbol{d}_i'. \tag{4.10}$$

Using the trace of a matrix, (4.10) can be rewritten to obtain

$$\operatorname{tr}\left(\sum_{i=1}^{N} \boldsymbol{R}\boldsymbol{d}_i'\boldsymbol{m}_i'^T\right) = \operatorname{tr}\left(\boldsymbol{R}\boldsymbol{H}\right), \tag{4.11}$$

With $\boldsymbol{H}$ defined as in (4.9). Now we have to find the matrix $\boldsymbol{R}$ that maximizes $\operatorname{tr}(\boldsymbol{R}\boldsymbol{H})$.

Lemma 4. *For all positive definite matrices $\boldsymbol{A}\boldsymbol{A}^T$ and every orthonormal matrix $\boldsymbol{B}$ it holds:*

$$tr\left(\boldsymbol{A}\boldsymbol{A}^T\right) \geq tr\left(\boldsymbol{B}\boldsymbol{A}\boldsymbol{A}^T\right).$$

Proof. Assume $\boldsymbol{a}_i$ to be the i-th column of $\boldsymbol{A}$. With that we can calculate

$$\begin{aligned}\operatorname{tr}\left(\boldsymbol{B}\boldsymbol{A}\boldsymbol{A}^T\right) &= \operatorname{tr}\left(\boldsymbol{A}^T\boldsymbol{B}\boldsymbol{A}\right) \\ &= \sum_{i=1}^{N} \boldsymbol{a}_i^T(\boldsymbol{B}\boldsymbol{a}_i^T).\end{aligned}$$

Using the inequality of Cauchy-Schwarz we derive:

$$\boldsymbol{a}_i^T(\boldsymbol{B}\boldsymbol{a}_i) \leq \sqrt{(\boldsymbol{a}_i^T\boldsymbol{a}_i)(\boldsymbol{a}_i^T\boldsymbol{B}^T\boldsymbol{B}\boldsymbol{a}_i)} = \boldsymbol{a}_i^T\boldsymbol{a}_i.$$

With the trace of the matrix we have

$$\operatorname{tr}\left(\boldsymbol{B}\boldsymbol{A}\boldsymbol{A}^T\right) \leq \sum_{i}^{N} \boldsymbol{a}_i^T\boldsymbol{a}_i = \operatorname{tr}\left(\boldsymbol{A}\boldsymbol{A}^T\right)$$

and therefore verified the claim of the lemma. □

Assume that the singular value decomposition of $\boldsymbol{H}$ is

$$\boldsymbol{H} = \boldsymbol{U}\boldsymbol{\Lambda}\boldsymbol{V}^T,$$

where $\boldsymbol{U}$ and $\boldsymbol{V}$ are orthonormal 3×3 matrices and $\boldsymbol{\Lambda}$ is a 3×3 diagonal matrix without negative elements. Suppose

$$\boldsymbol{R} = \boldsymbol{V}\boldsymbol{U}^T.$$

$\boldsymbol{R}$ is orthonormal and

$$\begin{aligned}\boldsymbol{RH} &= \boldsymbol{VU}^T\boldsymbol{U\Lambda V}^T\\ &= \boldsymbol{V\Lambda V}^T\end{aligned}$$

is a symmetric, positive definite matrix. Using the lemma above we have

$$\operatorname{tr}(\boldsymbol{RH}) \geq \operatorname{tr}(\boldsymbol{BRH})$$

for any orthonormal matrix $\boldsymbol{B}$. Therefore the matrix $\boldsymbol{R}$ is optimal. □

Finally, the optimal translation is calculated as (see Eq. (4.5) and (4.7))

$$\boldsymbol{t} = \boldsymbol{c}_m - \boldsymbol{Rc}_d.$$

Computing the Tranformation using Orthonormal Matrices

This algorithm is similar to the previous method and was independently developed 1988 by Horn, Hilden and Negahdaripour [62]. Again a correlation Matrix $\boldsymbol{H}$ according to Eq. (4.9) is calculated. Afterwards a so-called polar decomposition is computed, i.e., $\boldsymbol{H} = \boldsymbol{PS}$, where $\boldsymbol{S} = (\boldsymbol{H}^T\boldsymbol{H})^{1/2}$.

Theorem 2. *Given the matrices $\boldsymbol{H}$, $\boldsymbol{S}$ and $\boldsymbol{P}$ as described above. Then the optimal roation is given by*

$$\boldsymbol{R} = \boldsymbol{H}\left(\frac{1}{\sqrt{\lambda_1}}\boldsymbol{u}_1\boldsymbol{u}_1^T + \frac{1}{\sqrt{\lambda_2}}\boldsymbol{u}_2\boldsymbol{u}_2^T + \frac{1}{\sqrt{\lambda_3}}\boldsymbol{u}_3\boldsymbol{u}_3^T\right),$$

with $\{\lambda_i\}$ being the eigenvalue and $\{u_i\}$ the corresponding eigenvectors of the matrix $\boldsymbol{H}\boldsymbol{H}^T$ [62].

Proof. The rotation that minimizes the error in Eq. (4.2) is the orthonormal matrix that maximizes the trace of $\boldsymbol{RH}$ (see Eq. (4.11)). The algorithm must find the maximum and enforce the orthonormality constraint. The square matrix $\boldsymbol{H}$ can be decomposed into a product of an orthonormal matrix $\boldsymbol{P}$ and a positive semidefinite matrix $\boldsymbol{S}$ [62]. The matrix $\boldsymbol{S}$ is unambiguously defined. If $\boldsymbol{H}$ is not singular, then $\boldsymbol{P}$ is unique. Therefore, one can write:

$$\boldsymbol{H} = \boldsymbol{PS}, \tag{4.12}$$

where

$$\boldsymbol{S} = (\boldsymbol{H}^T\boldsymbol{H})^{1/2}$$

is the positive definite square root of the symmetric matrix $\boldsymbol{H}^T\boldsymbol{H}$ and

$$\boldsymbol{P} = \boldsymbol{H}(\boldsymbol{H}^T\boldsymbol{H})^{-1/2}$$

is an orthonormal matrix. It can be verified that $\boldsymbol{H} = \boldsymbol{PS}$, $\boldsymbol{S}^T = \boldsymbol{S}$ and $\boldsymbol{P}^T\boldsymbol{P} = \mathbb{1}$ hold (see Appendix A.3, Lemma 12).

Now we have to determine, how to compute the positive definite square root of a positive definite matrix. To this end, the matrix $\boldsymbol{H}^T\boldsymbol{H}$ is represented by its eigenvalues $\{\lambda_i\}$ and the set of corresponding eigenvectors $\{\boldsymbol{u}_i\}$ (see also Appendix A.3, Lemma 8):

$$\boldsymbol{H}^T\boldsymbol{H} = \lambda_1\boldsymbol{u}_1\boldsymbol{u}_1^T + \lambda_2\boldsymbol{u}_2\boldsymbol{u}_2^T + \lambda_3\boldsymbol{u}_3\boldsymbol{u}_3^T.$$

Since $\boldsymbol{H}^T\boldsymbol{H}$ is positive definite, the eigenvalues are positive and the root is a real number. The symmetric matrix $\boldsymbol{S}$ is constructed as follows:

$$\boldsymbol{S} = \sqrt{\lambda_1}\boldsymbol{u}_1\boldsymbol{u}_1^T + \sqrt{\lambda_2}\boldsymbol{u}_2\boldsymbol{u}_2^T + \sqrt{\lambda_3}\boldsymbol{u}_3\boldsymbol{u}_3^T.$$

The proof for the general case is given in Appendix A.3, lemma 10. In addition we have

$$\boldsymbol{S}^2 = \lambda_1\boldsymbol{u}_1\boldsymbol{u}_1^T + \lambda_2\boldsymbol{u}_2\boldsymbol{u}_2^T + \lambda_3\boldsymbol{u}_3\boldsymbol{u}_3^T$$

since eigenvectors are orthonormal and with all vectors $\boldsymbol{x}$ that are not the $\boldsymbol{0}$-vector we have

$$\boldsymbol{xSx} = \lambda_1(\boldsymbol{u}_1 \cdot \boldsymbol{x})^2 + \lambda_2(\boldsymbol{u}_2 \cdot \boldsymbol{x})^2 + \lambda_3(\boldsymbol{u}_3 \cdot \boldsymbol{x})^2 > 0.$$

Thus $\boldsymbol{S}$ is positive definite. The matrix $\boldsymbol{S}^{-1}$ is computed by the reciprocal values of the eigenvalues (cf. Appendix A.3, Corollary 3):

$$\boldsymbol{S}^{-1} = (\boldsymbol{H}^T\boldsymbol{H})^{-1/2} = \frac{1}{\sqrt{\lambda_1}}\boldsymbol{u}_1\boldsymbol{u}_1^T + \frac{1}{\sqrt{\lambda_2}}\boldsymbol{u}_2\boldsymbol{u}_2^T + \frac{1}{\sqrt{\lambda_3}}\boldsymbol{u}_3\boldsymbol{u}_3^T.$$

This term is used for setting up the orthonormal matrix

$$\boldsymbol{P} = \boldsymbol{HS}^{-1} = \boldsymbol{H}(\boldsymbol{H}^T\boldsymbol{H})^{-1/2}.$$

To determine the trace $\mathrm{tr}\,(\boldsymbol{RH})$ one uses the above method and yields

$$\begin{aligned} \mathrm{tr}\,(\boldsymbol{RH}) &= \mathrm{tr}\,(\boldsymbol{RPS}) \\ &= \sqrt{\lambda_1}\,\mathrm{tr}\left(\boldsymbol{RPu}_1\boldsymbol{u}_1^T\right) + \sqrt{\lambda_2}\,\mathrm{tr}\left(\boldsymbol{RPu}_2\boldsymbol{u}_2^T\right) \\ &\quad + \sqrt{\lambda_3}\,\mathrm{tr}\left(\boldsymbol{RPu}_3\boldsymbol{u}_3^T\right). \end{aligned} \tag{4.13}$$

The following lemma is true for all matrices:

Lemma 5. *For all matrices $\boldsymbol{X}$ and $\boldsymbol{Y}$, for which the products $\boldsymbol{XY}$ and $\boldsymbol{YX}$ yield a square matrix, it holds $tr(\boldsymbol{XY}) = tr(\boldsymbol{YX})$.* □

For all summands in (4.13) holds:

$$\mathrm{tr}\left(\boldsymbol{RPu}_i\boldsymbol{u}_i^T\right) = \mathrm{tr}\left(\boldsymbol{u}_i^T\boldsymbol{R}^T\boldsymbol{Pu}_i\right) = \mathrm{tr}\left(\boldsymbol{R}^T\boldsymbol{u}_i \cdot \boldsymbol{Pu}_i\right) = \boldsymbol{R}^T\boldsymbol{u}_i \cdot \boldsymbol{Pu}_i.$$

Since the vectors $\boldsymbol{u}_i$ are unit vectors by construction and $\boldsymbol{P}$ and $\boldsymbol{R}$ are orthonormal matrices, we have $\boldsymbol{R}^T\boldsymbol{u}_i \cdot \boldsymbol{P}\boldsymbol{u}_i \leq 1$. Equality is given, if and only if $\boldsymbol{R}^T\boldsymbol{u}_i = \boldsymbol{P}\boldsymbol{u}_i$ (see Appendix A.3, Lemma 7). It follows, that

$$\mathrm{tr}\,(\boldsymbol{RPS}) \leq \sqrt{\lambda_1} + \sqrt{\lambda_2} + \sqrt{\lambda_3} = \mathrm{tr}\,(\boldsymbol{S})$$

holds. According to this the maximum of $\mathrm{tr}\,(\boldsymbol{RPS})$ is reached, if $\boldsymbol{R}^T\boldsymbol{P} = \mathbb{1}$ and accordingly $\boldsymbol{R} = \boldsymbol{P}$ (cf. Appendix A.3, Lemma 9 and Corollary 1). The orthonormal matrix that occurs in (4.12), is the rotation matrix and it holds:

$$\boldsymbol{R} = \boldsymbol{P} = \boldsymbol{H}\boldsymbol{S}^{-1} = \boldsymbol{H}(\boldsymbol{H}^T\boldsymbol{H})^{-1/2}. \qquad \square$$

Again, the optimal translation is calculated as (cf. Eq. (4.5) and (4.7))

$$\boldsymbol{t} = \boldsymbol{c}_m - \boldsymbol{R}\boldsymbol{c}_d.$$

Computing the Tranformation using Unit Quaternions

The transformation for the ICP algorithm can be calculated by a method that uses unit quaternions. This method was invented 1987 by Horn [61]. The unit quaternion is an extension of the complex numbers to three imaginary parts [55, 56] and describes a rotation of angle θ about the rotation axis $\boldsymbol{a} = (a_x, a_y, a_z)$. It is the 4-vector

$$\dot{\boldsymbol{q}} = \begin{pmatrix} q_0 \\ q_x \\ q_y \\ q_z \end{pmatrix} = \begin{pmatrix} \cos(\theta/2) \\ \sin(\theta/2)\, a_x \\ \sin(\theta/2)\, a_y \\ \sin(\theta/2)\, a_z \end{pmatrix}, \tag{4.14}$$

with $q_0 \geq 0$ and $q_0^2 + q_x^2 + q_y^2 + q_z^2 = 1$ alternatively $\dot{\boldsymbol{q}} \cdot \dot{\boldsymbol{q}} = 1$. Furthermore, every quaternion $\dot{\boldsymbol{q}}$ has a matrix representation $\boldsymbol{Q}$ and $\bar{\boldsymbol{Q}}$, where

$$\boldsymbol{Q} = \begin{pmatrix} q_0 & -q_x & -q_y & -q_z \\ q_x & q_0 & -q_z & q_y \\ q_y & q_z & q_0 & -q_x \\ q_z & -q_y & q_x & q_0 \end{pmatrix} \quad \text{and} \quad \bar{\boldsymbol{Q}} = \begin{pmatrix} q_0 & -q_x & -q_y & -q_z \\ q_x & q_0 & q_z & -q_y \\ q_y & -q_z & q_0 & q_x \\ q_z & q_y & -q_x & q_0 \end{pmatrix}. \tag{4.15}$$

A vector $\boldsymbol{v} = (v_x, v_y, v_z)^T \in \mathbb{R}^3$ is represented by a quaternion, which first component is zero, i.e., $\dot{\boldsymbol{v}} = (0, v_x, v_y, v_z)^T$. The rotated vector $\boldsymbol{v}_{\mathrm{rot}}$ is computed by a left-hand multiplication with the quaternion $\dot{\boldsymbol{q}}$ and the right-multiplication with the conjugate quaternion $\dot{\boldsymbol{q}}^*$. For unit quaternions conjugation equals inverting. For $\boldsymbol{v}_{\mathrm{rot}}$ it holds:

$$\dot{\boldsymbol{v}}_{\text{rot}} = \dot{\boldsymbol{q}}\dot{\boldsymbol{v}}\dot{\boldsymbol{q}}^* = (\boldsymbol{Q}\dot{\boldsymbol{v}})\dot{\boldsymbol{q}}^* = \boldsymbol{Q}^T(\boldsymbol{Q}\dot{\boldsymbol{v}}) = (\boldsymbol{Q}^T\boldsymbol{Q})\dot{\boldsymbol{v}}. \tag{4.16}$$

The 3×3 rotation matrix $\boldsymbol{R}$ is caluculated by the following schema from the unit quaternion [11]:

$$\boldsymbol{R} = \begin{pmatrix} p_0^2 + p_x^2 - p_y^2 - p_z^2 & 2p_xp_y - 2p_0p_z & 2p_xp_z + 2p_0p_y \\ 2p_0p_z + 2p_xp_y & p_0^2 - p_x^2 + p_y^2 - p_z^2 & 2p_yp_z - 2p_0p_x \\ 2p_xp_z - 2p_0p_y & 2p_0p_x + 2p_yp_z & p_0^2 - p_x^2 - p_y^2 + p_z^2 \end{pmatrix} \tag{4.17}$$

Theorem 3. *The rotation represented as unit quaternion $\dot{\boldsymbol{q}}$, that minimizes (4.1), corresponds to the largest eigenvalue of the cross covariance matrix*

$$\boldsymbol{N} = \begin{pmatrix} (S_{xx} + S_{yy} + S_{zz}) & (S_{yz} + S_{zy}) & (S_{zx} + S_{xz}) & (S_{xy} + S_{yx}) \\ (S_{yz} + S_{zy}) & (S_{xx} - S_{yy} - S_{zz}) & (S_{xy} + S_{yx}) & (S_{zx} + S_{xz}) \\ (S_{zx} + S_{xz}) & (S_{xy} + S_{yx}) & (-S_{xx} + S_{yy} - S_{zz}) & (S_{yz} + S_{zy}) \\ (S_{xy} + S_{yx}) & (S_{yz} + S_{zy}) & (S_{zx} + S_{xz}) & (-S_{xx} - S_{yy} + S_{zz}) \end{pmatrix},$$

where $S_{xx} = \sum_{i=1}^{N_m} m'_{x,i} d'_{x,i}$, $S_{xy} = \sum_{i=1}^{N} m'_{x,i} d'_{iy}$, $\ldots$.

Proof. In line with the argumentation of the previous methods we note that it is sufficient to maximize the sum (4.10). Now the quaternions come into play: To find the rotation matrix $\boldsymbol{R}$ equals finding the unit quaternion $\dot{\boldsymbol{q}}$ that maximizes the term

$$\sum_{i=1}^{N} \dot{\boldsymbol{m}}'_i \cdot (\dot{\boldsymbol{q}}\dot{\boldsymbol{d}}'_i\dot{\boldsymbol{q}}^*) = \sum_{i=1}^{N} (\dot{\boldsymbol{q}}\dot{\boldsymbol{m}}'_i) \cdot (\dot{\boldsymbol{d}}'_i\dot{\boldsymbol{q}})$$

(see Eq. (4.16)). Suppose $\bar{\boldsymbol{M}}_i$ and $\boldsymbol{D}_i$ are now the matrices that correspond to the quaternions $\dot{\boldsymbol{m}}'_i$ and $\dot{\boldsymbol{d}}'_i$ as given by Eq. (4.15). According to (4.16) the sum that has to be maximized is

$$\sum_{i=1}^{N} (\bar{\boldsymbol{M}}_i\dot{\boldsymbol{q}}) \cdot (\boldsymbol{D}_i\dot{\boldsymbol{q}})$$

or

$$\sum_{i=1}^{N} \dot{\boldsymbol{q}}^T \bar{\boldsymbol{M}}_i \boldsymbol{D}_i^T \dot{\boldsymbol{q}} = \dot{\boldsymbol{q}}^T \left(\sum_{i=1}^{N} \bar{\boldsymbol{M}}_i \boldsymbol{D}_i^T \right) \dot{\boldsymbol{q}}.$$

After writing the matrices for $\bar{\boldsymbol{M}}_i$ and $\boldsymbol{D}_i$, a calculation yields:

$$\dot{\boldsymbol{q}}^T \left(\sum_{i=1}^{N} \bar{\boldsymbol{M}}_i \boldsymbol{D}_i^T \right) \dot{\boldsymbol{q}} = \dot{\boldsymbol{q}}^T \left(\sum_{i=1}^{N} \begin{pmatrix} 0 & -m'_{x,i} & -m'_{y,i} & -m'_{z,i} \\ m'_{x,i} & 0 & m'_{z,i} & -m'_{y,i} \\ m'_{y,i} & -m'_{z,i} & 0 & m'_{x,i} \\ m'_{z,i} & m'_{y,i} & -m'_{x,i} & 0 \end{pmatrix} \begin{pmatrix} 0 & -d'_{x,i} & -d'_{y,i} & -d'_{z,i} \\ d'_{x,i} & 0 & -d'_{z,i} & d'_{y,i} \\ d'_{y,i} & d'_{z,i} & 0 & -d'_{x,i} \\ d'_{z,i} & -d'_{y,i} & d'_{x,i} & 0 \end{pmatrix}^T \right) \dot{\boldsymbol{q}}$$

$$= \dot{\boldsymbol{q}}^T \boldsymbol{N} \dot{\boldsymbol{q}}. \tag{4.18}$$

The matrix $\boldsymbol{N}$ is the result of the multiplication. The following lemma shows how the unit quaternion that maximizes (4.15) is calculated:

Lemma 6. *The unit quaternion $\dot{\boldsymbol{q}}$ (it holds $\dot{\boldsymbol{q}} \cdot \dot{\boldsymbol{q}} = 1$), that maximizes the term $\dot{\boldsymbol{q}}^T \boldsymbol{N} \dot{\boldsymbol{q}}$ (see Eq. (4.18)) is the eigenvector with the largest eigenvalue of the matrix $\boldsymbol{N}$ [61].*

Proof. The symmetric matrix $\boldsymbol{N}$ is a 4×4 matrix. Therefore $\boldsymbol{N}$ has four eigenvalues, i.e., $(\lambda_1, \lambda_2, \lambda_3, \lambda_4)$. One can construct four eigenvectors $(\dot{\boldsymbol{e}}_1, \dot{\boldsymbol{e}}_2, \dot{\boldsymbol{e}}_3, \dot{\boldsymbol{e}}_4)$ corresponding to the eigenvalues, such that

$$\boldsymbol{N} \dot{\boldsymbol{e}}_i = \lambda_i \dot{\boldsymbol{e}}_i \qquad \text{for} \quad i = 1, 2, 3, 4 \qquad \text{holds.}$$

The eigenvectors span a four-dimensional vector space, i.e., they are linearly independent. Thus any quaternion $\dot{\boldsymbol{q}}$ is writable as linear combination

$$\dot{\boldsymbol{q}} = \alpha_1 \dot{\boldsymbol{e}}_1 + \alpha_2 \dot{\boldsymbol{e}}_2 + \alpha_3 \dot{\boldsymbol{e}}_3 + \alpha_4 \dot{\boldsymbol{e}}_4.$$

Since eigenvectors are orthogonal, we have:

$$\dot{\boldsymbol{q}} \cdot \dot{\boldsymbol{q}} = \alpha_1^2 + \alpha_2^2 + \alpha_3^2 + \alpha_4^2.$$

This equation must equal 1, since we search for the unit quaternion $\dot{\boldsymbol{q}}$. Furthermore, we derive

$$\boldsymbol{N} \dot{\boldsymbol{q}} = \alpha_1 \lambda_1 \dot{\boldsymbol{e}}_1 + \alpha_2 \lambda_2 \dot{\boldsymbol{e}}_2 + \alpha_3 \lambda_3 \dot{\boldsymbol{e}}_3 + \alpha_4 \lambda_4 \dot{\boldsymbol{e}}_4,$$

since $\dot{\boldsymbol{e}}_1, \dot{\boldsymbol{e}}_2, \dot{\boldsymbol{e}}_3, \dot{\boldsymbol{e}}_4$ are eigenvectors of $\boldsymbol{N}$. We conclude, that

$$\dot{\boldsymbol{q}}^T \boldsymbol{N} \dot{\boldsymbol{q}} = \dot{\boldsymbol{q}} \cdot (\boldsymbol{N} \dot{\boldsymbol{q}}) = \alpha_1^2 \lambda_1 + \alpha_2^2 \lambda_2 + \alpha_3^2 \lambda_3 + \alpha_4^2 \lambda_4$$

holds. Assume the eigenvalues are sorted according to their values, i.e., $\lambda_1 \geq \lambda_2 \geq \lambda_3 \geq \lambda_4$, then the following inequality holds true:

$$\dot{\boldsymbol{q}}^T \boldsymbol{N} \dot{\boldsymbol{q}} \leq \alpha_1^2 \lambda_1 + \alpha_2^2 \lambda_1 + \alpha_3^2 \lambda_1 + \alpha_4^2 \lambda_1 = \lambda_1.$$

This shows that the quadratic form is never larger than the largest eigenvalue. Choosing $\alpha_1 = 1$ and $\alpha_2 = \alpha_3 = \alpha_4 = 0$ the maximal value is reached and the unit quaternion is $\dot{\boldsymbol{q}} = \dot{\boldsymbol{e}}_1$ [61]. □

After the computation of the rotation matrix $\boldsymbol{R}$ the translation is

$$\boldsymbol{t} = \boldsymbol{c}_m - \boldsymbol{R}\boldsymbol{c}_d.$$

Computing the Tranformation using Dual Quaterions

The transformation that minimizes Eq. (4.1) can be computed by an algorithm that uses so-called dual quaternions. This method was developed by Walker, Shao and Volz in 1991 [132]. Unlike the first three methods covered so far the transformation is found by one step. There is no need to apply the trick with centroids to compute the rotation in a separate fashion.

The quaternion $\dot{\boldsymbol{q}}$ (cf. Eq. (4.14)) is the 4-vector $(q_0, q_x, q_y, q_z)^T$ or the pair $(q_0, \boldsymbol{q})^T$. A dual quaternion $\hat{\dot{\boldsymbol{q}}}$ consists of two quaternions $\dot{\boldsymbol{q}}$ and $\dot{\boldsymbol{s}}$, i.e.,

$$\hat{\dot{\boldsymbol{q}}} = \dot{\boldsymbol{q}} + \varepsilon \dot{\boldsymbol{s}},$$

where a special multiplication rule for ε is defined: $\varepsilon^2 = 0$. In addition to this, the matrix notation of the quaterion is important (see Eq. (4.15)):

$$\boldsymbol{Q} = \begin{pmatrix} q_0 & -q_x & -q_y & -q_z \\ q_x & q_0 & -q_z & q_y \\ q_y & q_z & q_0 & -q_x \\ q_z & -q_y & q_x & q_0 \end{pmatrix} = \begin{pmatrix} q_0 & -\boldsymbol{q}^T \\ \boldsymbol{q} & q_0 \mathbb{1}_{3\times 3} + \boldsymbol{C}_{\boldsymbol{q}} \end{pmatrix}, \tag{4.19}$$

$$\bar{\boldsymbol{Q}} = \begin{pmatrix} q_0 & -q_x & -q_y & -q_z \\ q_x & q_0 & q_z & -q_y \\ q_y & -q_z & q_0 & q_x \\ q_z & q_y & -q_x & q_0 \end{pmatrix} = \begin{pmatrix} q_0 & -\boldsymbol{q}^T \\ \boldsymbol{q} & q_0 \mathbb{1}_{3\times 3} - \boldsymbol{C}_{\boldsymbol{q}} \end{pmatrix}. \tag{4.20}$$

The matrix $\boldsymbol{C}_{\boldsymbol{q}}$ is defined as follows:

$$\boldsymbol{C}_{\boldsymbol{q}} = \begin{pmatrix} 0 & -q_z & q_y \\ q_z & 0 & -q_x \\ -q_y & q_x & 0 \end{pmatrix}. \tag{4.21}$$

The rigid 3D transformation $(\boldsymbol{R}, \boldsymbol{t})$ meets the following conditions

$$\dot{\boldsymbol{q}} \cdot \dot{\boldsymbol{q}} = 1 \qquad \text{and} \qquad \dot{\boldsymbol{q}} \cdot \dot{\boldsymbol{s}} = 0. \tag{4.22}$$

The rotation matrix $\boldsymbol{R}$ is written as

$$\boldsymbol{R} = (q_0^2 - \boldsymbol{q}^T\boldsymbol{q})\mathbb{1} + 2\boldsymbol{q}\boldsymbol{q}^T + 2q_0\boldsymbol{C}_{\boldsymbol{q}} \tag{4.23}$$

which is equivalent to the representation in Eq. (4.17). The translation vector is $\boldsymbol{t} = \boldsymbol{p}$, where $\boldsymbol{p}$ is the vector part of the quaternion $\dot{\boldsymbol{p}}$ that is given by:

$$\dot{\boldsymbol{p}} = \bar{\boldsymbol{Q}}^T\dot{\boldsymbol{s}} \tag{4.24}$$

The first entry p_0 of $\dot{\boldsymbol{p}}$ is always zero. The 3-vector $\boldsymbol{v} = (v_x, v_y, v_z)^T \in \mathbb{R}^3$ is again represented as quaternion $\dot{\boldsymbol{v}} = (0, v_x, v_y, v_z)^T$, with the first component being equal to zero. With these definitions it is easy to show the equality:

$$\boldsymbol{R}\boldsymbol{x} + \boldsymbol{t} = \bar{\boldsymbol{Q}}^T\boldsymbol{Q}\dot{\boldsymbol{x}} + \bar{\boldsymbol{Q}}^T\dot{\boldsymbol{s}}.$$

Theorem 4. *The transformation consisting of rotation and translation that minimized the error function Eq. (4.1) is the solution of the eigenvalue problem of a 4×4 matrix function that is built from corresponding point pairs.*

Proof. Using the definitions of the dual quaternion from above the error function (4.2) is rewritten as:

$$E(\boldsymbol{R}, \boldsymbol{t}) = E(\dot{\boldsymbol{q}}, \dot{\boldsymbol{s}}) = \frac{1}{N}\left[\dot{\boldsymbol{q}}^T\boldsymbol{C}_1\dot{\boldsymbol{q}} + N\dot{\boldsymbol{s}}^T\dot{\boldsymbol{s}} + \dot{\boldsymbol{s}}^T\boldsymbol{C}_2\dot{\boldsymbol{q}} + \text{const.}\right] ,$$

where the terms $\boldsymbol{C}_1, \boldsymbol{C}_2$, and const. are defined using the quaternion representation $\boldsymbol{M}_i$ and $\boldsymbol{D}_i$, and accordingly $\boldsymbol{C}_{\boldsymbol{m}_i}$ and $\boldsymbol{C}_{\boldsymbol{d}_i}$ for the 3D data points $\boldsymbol{m}_i$ and $\boldsymbol{d}_i$ (cf. Eq. (4.19), (4.20) and (4.21)) as:

$$\boldsymbol{C}_1 = -2\sum_{i=1}^{N}\boldsymbol{M}_i^T\bar{\boldsymbol{D}}_i = -2\sum_{i=1}^{N}\begin{pmatrix} \boldsymbol{C}_{\boldsymbol{m}_i}\boldsymbol{C}_{\boldsymbol{d}_i} + \boldsymbol{m}_i\boldsymbol{d}_i^T & -\boldsymbol{C}_{\boldsymbol{m}_i}\boldsymbol{d}_i \\ -\boldsymbol{m}_i^T\boldsymbol{C}_{\boldsymbol{d}_i} & \boldsymbol{m}_i^T\boldsymbol{d}_i \end{pmatrix},$$

$$\boldsymbol{C}_2 = -2\sum_{i=1}^{N}\bar{\boldsymbol{D}}_i - \boldsymbol{M}_i = -2\sum_{i=1}^{N}\begin{pmatrix} -\boldsymbol{C}_{\boldsymbol{d}_i} - \boldsymbol{C}_{\boldsymbol{m}_i} & \boldsymbol{d}_i - \boldsymbol{m}_i \\ -(\boldsymbol{d}_i - \boldsymbol{m}_i)^T & 0 \end{pmatrix},$$

$$\text{const.} = \sum_{i=1}^{N}(\boldsymbol{d}_i^T\boldsymbol{d}_i + \boldsymbol{m}_i^T\boldsymbol{m}_i).$$

After adding the terms, the condition for the transformation, i.e., Eq. (4.22), the error function becomes

$$\begin{aligned} E(\dot{\boldsymbol{q}}, \dot{\boldsymbol{s}}) = &\frac{1}{N}\left[\dot{\boldsymbol{q}}^T\boldsymbol{C}_1\dot{\boldsymbol{q}} + N\dot{\boldsymbol{s}}^T\dot{\boldsymbol{s}} + \dot{\boldsymbol{s}}^T\boldsymbol{C}_2\dot{\boldsymbol{q}} + \text{const.}\right. \\ &\left. + \lambda_1(\dot{\boldsymbol{q}}^T\dot{\boldsymbol{q}} - 1) + \lambda_2(\dot{\boldsymbol{s}}^T\dot{\boldsymbol{q}})\right] , \end{aligned} \tag{4.25}$$

here λ_1 and λ_2 are Lagrange multipliers (see Appendix A.5). To determine the minimum of the function the partial derivatives are calculated and set to zero:

$$\frac{\partial E(\dot{\boldsymbol{q}}, \dot{\boldsymbol{s}})}{\partial \dot{\boldsymbol{q}}} = \frac{1}{N}\left[(\boldsymbol{C}_1 + \boldsymbol{C}_1^T)\dot{\boldsymbol{q}} + \boldsymbol{C}_2^T \dot{\boldsymbol{s}} + 2\lambda_1 \dot{\boldsymbol{q}} + \lambda_2 \dot{\boldsymbol{s}}\right] = 0 \tag{4.26}$$

$$\frac{\partial E(\dot{\boldsymbol{q}}, \dot{\boldsymbol{s}})}{\partial \dot{\boldsymbol{q}}} = \frac{1}{N}\left[2N\dot{\boldsymbol{s}} + \boldsymbol{C}_2 \dot{\boldsymbol{q}} + \lambda_2 \dot{\boldsymbol{q}}\right] = 0. \tag{4.27}$$

Multiplying Eq. (4.27) with $\dot{\boldsymbol{q}}$ yields $\lambda_2 = -\dot{\boldsymbol{q}}^T \boldsymbol{C}_2 \dot{\boldsymbol{q}} = 0$, because the matrix $\boldsymbol{C}_2$ is skew symmetric. Thus $\dot{\boldsymbol{s}}$ is given by:

$$\dot{\boldsymbol{s}} = -\frac{1}{2N}\boldsymbol{C}_2 \dot{\boldsymbol{q}}. \tag{4.28}$$

Using (4.28) in Eq. (4.26), we have

$$\boldsymbol{A}\dot{\boldsymbol{q}} = \lambda_1 \dot{\boldsymbol{q}}, \qquad \text{where} \qquad \boldsymbol{A} = \frac{1}{2}\left(\frac{1}{2N}\boldsymbol{C}_2^T \boldsymbol{C}_2 - \boldsymbol{C}_1 - \boldsymbol{C}_1^T\right). \tag{4.29}$$

Therefore, the quaternion $\dot{\boldsymbol{q}}$ is the eigenvector of the matrix $\boldsymbol{A}$ and λ_1 is the corresponding eigenvalue. Substituting (4.29) back into Eq. (4.25) results in

$$E(\dot{\boldsymbol{q}}, \dot{\boldsymbol{s}}) = \frac{1}{N}(\text{const.} - \lambda_1).$$

Hence the error is minimal, if the eigenvector with the largest eigenvalue is chosen.

After the quaternion $\dot{\boldsymbol{q}}$ is calculated the quaternion $\dot{\boldsymbol{s}}$ is computed according to (4.28). The translation and rotation is then given by the formulas (4.23) and (4.24). □

Lorusso et al. have firstly implemented and evaluated these four methods [76]. All these methods show similar computational requirements ($\mathcal{O}$(number of point pairs) with similar constants) and have all about the same accuracies depending on the resolution and similar stabilities with respect to noise data.

The Minimal Number of Corresponding Points

The closed form methods for computing the transformation (rotation $\boldsymbol{R}$ and translation $\boldsymbol{t}$) for the ICP algorithm based on the corresponding points perform a low number of operations. Due to the orthonormality constraint the 9 entries of the rotation matrix contain only three free parameters: the rotation about the x-, y- and z- axis. Furthermore, three free parameters for the translation have to be computed. Point pairs in $\mathbb{R}^3$ induce 3 equations and therefore two pairs are necessary to compute the optimal transformation $(\boldsymbol{R}, \boldsymbol{t})$. Because two points are always co-linear, the rotation is not determined unambiguously and an additional point pair is needed [61].

Fig. 4.1 Visualisation of the iterations of the ICP algorithm. The initial attitude (left) of two 3D scans, the attitude after three iterations (middle) and after 15 iterations (right) is presented. In all steps a rotation $\boldsymbol{R}$ and a translation $\boldsymbol{t}$ is computed in closed form and applied to the second scan.

Figure 4.1 shows three steps of the ICP algorithm. The computed transformation is applied to the second scan.

4.1.2 Approximate Solution of the ICP Error Function by a Helical Motion

Under the assumption the transformation $(\boldsymbol{R}, \boldsymbol{t})$ that has to be calculated by the ICP algorithm is small we can approximate the solution by applying instantaneous kinematics. This solution was initially given by Hofer and Pottmann [59, 98] and first applied to registration tasks in robotics by Rusu et al. [108].

Instantaneous kinematics computes a displacements of a 3D point by an affine transformation via a so-called helical motion [98]. A 3D point $\boldsymbol{x}$ is mapped by adding a small displacement $\boldsymbol{v}(\boldsymbol{x})$, i.e.,

$$\boldsymbol{v}(\boldsymbol{x}) = \bar{\boldsymbol{c}} + \boldsymbol{c} \times \boldsymbol{x}$$

To minimize the ICP error function Eq. (4.1) the displacement of the points in D has to minimize the distances between the point pairs. Thus Eq. (4.2), is rewritten as follows:

$$\begin{aligned} E(\bar{\boldsymbol{c}}, \boldsymbol{c}) &= \sum_{i=1}^{N} ||\boldsymbol{m}_i - (\boldsymbol{d}_i + \boldsymbol{v}(d_i))||^2 \\ &= \sum_{i=1}^{N} ||\boldsymbol{m}_i - (\boldsymbol{d}_i + \bar{\boldsymbol{c}} + \boldsymbol{c} \times \boldsymbol{d}_i)||^2 \end{aligned} \quad (4.30)$$

Since Eq. (4.30) is a quadratic function with the two unknown variables $\bar{\boldsymbol{c}}, \boldsymbol{c}$ the following theorem holds:

Theorem 5. *Suppose the vectors* $\boldsymbol{a}_i$ *and* $\boldsymbol{u}$ *to be* $\boldsymbol{a}_i = \boldsymbol{m}_i - \boldsymbol{d}_i$ *and* $\boldsymbol{u} = (\bar{\mathbf{c}}^T, \mathbf{c}^T)$. *Then the optimal displacement for Eq. (4.30) is given by the solution to the following linear equation system:* $\bar{\mathbf{c}} + \mathbf{c} \times \boldsymbol{d}_i = \boldsymbol{B}(\boldsymbol{d}_i) \cdot \boldsymbol{u}$ *with*

$$\boldsymbol{B}(\boldsymbol{d}_i) = \begin{pmatrix} 0 & d_{z,i} & -d_{y,i} & 1 & 0 & 0 \\ -d_{z,i} & 0 & d_{x,i} & 0 & 1 & 0 \\ d_{y,i} & -d_{x,i} & 0 & 0 & 0 & 1 \end{pmatrix}$$

$$\sum_{i=1}^{N} \boldsymbol{B}(\boldsymbol{d}_i)^T \boldsymbol{B}(\boldsymbol{d}_i)\boldsymbol{u} = \sum_{i=1}^{N} \boldsymbol{B}(\boldsymbol{d}_i)\boldsymbol{a}_i \tag{4.31}$$

Proof. Proving the theorem is straightforward, since $E(\bar{\mathbf{c}}, \mathbf{c})$ is a quadratic function. □

The optimal displacement calculated by Eq. (4.31) corresponds to an affine motion. Therefore in a post processing step a rigid transformation $(\boldsymbol{R}, \boldsymbol{t})$ is calculated from $(\bar{\mathbf{c}}, \mathbf{c})$. Figure 4.2 presents the displacement of a point using the affine transformation and rigid transformation.

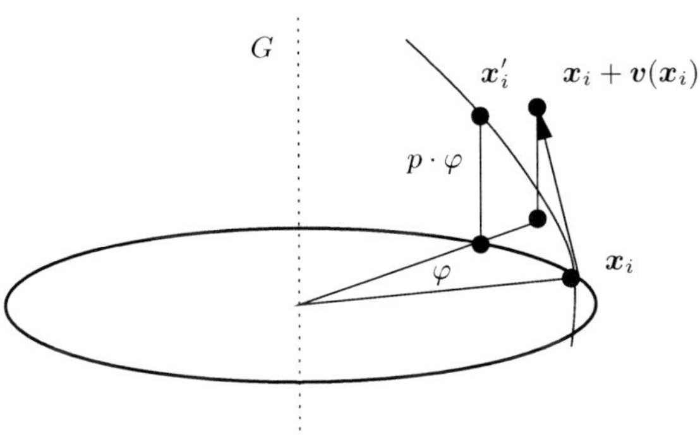

Fig. 4.2 The affine position of a 3D point $\boldsymbol{x}_i + \boldsymbol{v}(\boldsymbol{x}_i)$ is different from the rigid transformation that results in point $\boldsymbol{x}_i'$.. Based on [59].

The rigid transformation is calculated as follows: If $\mathbf{c} = \mathbf{0}$ only a translation is present. In this case it holds $\boldsymbol{t} = \bar{\mathbf{c}}$. Otherwise an axis G consisting of a direction vector $\boldsymbol{g}$ and a momentum vector $\bar{\boldsymbol{g}}$ can be computed. The tuple $(\boldsymbol{g}, \bar{\boldsymbol{g}})$ are the Plücker coordinates of the axis G (cf. Appendix A.2):

$$\boldsymbol{g} = \frac{\mathbf{c}}{||\mathbf{c}||}, \quad \bar{\boldsymbol{g}} = \frac{\mathbf{c} - p\mathbf{c}}{||\mathbf{c}||}, \quad p = \frac{\mathbf{c}^T \cdot \bar{\mathbf{c}}}{\mathbf{c}^2}$$

Based on G the point $\boldsymbol{d}_i$ has to be transformed as follows:

$$\boldsymbol{d}'_i = \boldsymbol{R}(\boldsymbol{d}_i - \boldsymbol{p}) + (p\ \varphi)\boldsymbol{g} + \boldsymbol{p} \tag{4.32}$$

Here $\boldsymbol{R}$ is the rotation matrix

$$\boldsymbol{R} = \frac{1}{b_0^2 + b_1^2 + b_2^2 + b_3^2} \begin{pmatrix} b_0^2 + b_1^2 - b_2^2 - b_3^2 & 2(b_1 b_2 + b_0 b_3) & 2(b_1 b_3 - b_0 b_2) \\ 2(b_1 b_2 - b_0 b_3) & b_0^2 - b_1^2 + b_2^2 - b_3^2 & 2(b_2 b_3 + b_0 b_1) \\ 2(b_1 b_3 + b_0 b_2) & 2(b_2 b_3 - b_0 b_1) & b_0^2 - b_1^2 - b_2^2 + b_3^2 \end{pmatrix} \tag{4.33}$$

where $b_0 = \cos(\varphi/2)$, $b_1 = g_x \sin(\varphi/2)$, $b_2 = g_y \sin(\varphi/2)$, and $b_3 = g_z \sin(\varphi/2)$. $\boldsymbol{g} = (g_x, g_y, g_z)$ is the above mentioned direction vector of the axis G. Furthermore in Eq. (4.32) $\boldsymbol{p}$ is an arbitrary point on the axis G. Note: Eq. (4.33) is similar to (4.17). Please refer to [59] for more details.

Since the minimization of the ICP error function using the helical motion is an approximative solution, one will need more iterations for convergence of the ICP algorithm.

4.1.3 Linearized Solution of the ICP Error Function

As in the last section, we linearize the rotation. Given a rotation matrix based on the Euler angles (cf. Apendix A.1.1)

$$\boldsymbol{R} = \begin{pmatrix} \cos\theta_y \cos\theta_z & -\cos\theta_y \sin\theta_z & \sin\theta_y \\ \cos\theta_z \sin\theta_x \sin\theta_y + \cos\theta_x \sin\theta_z & \cos\theta_x \cos\theta_z - \sin\theta_x \sin\theta_y \sin\theta_z & -\cos\theta_y \sin\theta_x \\ \sin\theta_x \sin\theta_z - \cos\theta_x \cos\theta_z \sin\theta_y & \cos\theta_z \sin\theta_x + \cos\theta_x \sin\theta_y \sin\theta_z & \cos\theta_x \cos\theta_y \end{pmatrix}. \tag{4.34}$$

we use the first-order Taylor series approximation that is valid for small angles

$$\sin\theta \approx \theta - \frac{\theta^3}{3} + \frac{\theta^5}{5} - \cdots$$
$$\cos\theta \approx 1 - \frac{\theta^2}{2} + \frac{\theta^4}{4} - \cdots$$

and apply is to (4.34). The resulting approximativ rotation is

$$\boldsymbol{R} \approx \begin{pmatrix} 1 & -\theta_z & \theta_y \\ \theta_x\theta_y + \theta_z & 1 - \theta_x\theta_y\theta_z & -\theta_x \\ \theta_x\theta_z - \theta_y & \theta_x + \theta_y\theta_z & 1 \end{pmatrix}. \tag{4.35}$$

As a second approximation we assume that the result of a multiplication of small angles yields even smaller values that can be omitted as well. This eliminates second order and higher combination terms and Eq. (4.35) becomes:

$$\boldsymbol{R} \approx \begin{pmatrix} 1 & -\theta_z & \theta_y \\ \theta_z & 1 & -\theta_x \\ \theta_y & \theta_x & 1 \end{pmatrix} = \mathbb{1}_{3\times 3} - \begin{pmatrix} 0 & -\theta_z & \theta_y \\ \theta_z & 0 & -\theta_x \\ \theta_y & \theta_x & 0 \end{pmatrix}. \tag{4.36}$$

We notice, that this term is closely related to the Rodrigues formula, but no longer orthonormal.

Replacing this approximation (4.36) in the ICP error function Eq. (4.2) and rearranging the unknown variables in a vector yields:

$$\begin{aligned} \boldsymbol{m}_i - (\boldsymbol{R}\boldsymbol{d}_i + \boldsymbol{t}) &\approx \boldsymbol{m}_i - \begin{pmatrix} 1 & -\theta_z & \theta_y \\ \theta_z & 1 & -\theta_x \\ \theta_y & \theta_x & 1 \end{pmatrix} \boldsymbol{d}_i - \boldsymbol{t} \\ &= \boldsymbol{m}_i - \boldsymbol{d}_i - \begin{pmatrix} 0 & d_{z,i} & -d_{y,i} & 1 & 0 & 0 \\ -d_{z,i} & 0 & d_{x,i} & 0 & 1 & 0 \\ d_{y,i} & -d_{x,i} & 0 & 0 & 0 & 1 \end{pmatrix} \begin{pmatrix} \theta_z \\ \theta_y \\ \theta_x \\ t_x \\ t_y \\ t_z \end{pmatrix}. \end{aligned} \tag{4.37}$$

Surprisingly, the resulting equation equals the notation of Eq. (4.30) by Hofer and Pottmann [59,98]. The top part of the solution vector was called **c** while the bottom part resambles $\bar{\mathbf{c}}$. Therefore, the result has to be interpreted as a so-called helical motion (cf. Section 4.1.2) and not using the small angle assumption. The small angly approxmation fails, since the rotation calculated by the ICP algorithm refers always to the global coordinate system, i.e., represents a rotation about the origin $(0, 0, 0)$. While the helical motion takes care of this by calculating a rotation axis, or approximation does not regard this.

In order to make the small angle approximation of the rotation matrix Eq. (4.36) to work, we have to apply the centroid trick, already used to derive Eq. (4.8). This separates the rotation from the translation. The resulting term is

$$\boldsymbol{m}'_i - \boldsymbol{R}\boldsymbol{d}'_i \approx \boldsymbol{m}'_i - \boldsymbol{d}'_i - \begin{pmatrix} 0 & d_{z,i} & -d_{y,i} \\ -d_{z,i} & 0 & d_{x,i} \\ d_{y,i} & -d_{x,i} & 0 \end{pmatrix} \begin{pmatrix} \theta_z \\ \theta_y \\ \theta_x \end{pmatrix}$$

Theorem 6. *Suppose the vectors* $\boldsymbol{a}_i$ *and* $\boldsymbol{u}$ *to be* $\boldsymbol{a}_i = \boldsymbol{m}_i - \boldsymbol{d}_i$ *and* $\boldsymbol{u} = (\theta_z, \theta_y, \theta_x))$. *Then the optimal displacement for the ICP errorfunction composed using the approximation in Eq.* (4.37) *is given by the solution to the following linear equation system:*

$$\boldsymbol{D}(\boldsymbol{d}_i) = \begin{pmatrix} 0 & d_{z,i} & -d_{y,i} \\ -d_{z,i} & 0 & d_{x,i} \\ d_{y,i} & -d_{x,i} & 0 \end{pmatrix}$$

$$\sum_{i=1}^{N} \boldsymbol{D}(\boldsymbol{d}_i)^T \boldsymbol{D}(\boldsymbol{d}_i)\boldsymbol{u} = \sum_{i=1}^{N} \boldsymbol{D}(\boldsymbol{d}_i)\boldsymbol{a}_i \tag{4.38}$$

Proof. Proving the theorem is again straightforward, since the term (4.38) is quadratic. □

After the approximation of the roation is foud, the optimal translation is calculated as (see Eq. (4.5) and (4.7))

$$\boldsymbol{t} = \boldsymbol{c}_m - \boldsymbol{R}\boldsymbol{c}_d.$$

4.1.4 Computing Closest Points

The crucial step of the ICP scan matching is the computation of closest points. To reduce the computing time of the ICP algorithm, the determination of closest points in the set M, given a point $\boldsymbol{p}_q \in D$, must be accelerated. A naïve implementation examines all points in M and the resulting computation time for ICP is $\mathcal{O}(|N_D|\,|N_M|)$, i.e., $\mathcal{O}(n^2)$. Different approaches have been proposed to solve the closest point problem faster. Technically the problem is described as follows: The search space contains N entries, i.e., the points, that are described by k real-valued keys. In addition we have a distance measure, the Euclidean distance.

Binary Search. If $k = 1$ a binary search returns in $\mathcal{O}(\log n)$ worst case time the closest points. The entries have to be sorted first by any sorting algorithm.

General Voronoi Diagrams. If $k = 2$ a Voronoi diagram can be computed for the given set of points. A Voronoi diagram divides the space into cells such that these cells determine the closest point. This can be done in $\mathcal{O}(n)$ time. A point-cell-algorithm is used to localize the cell where the query $\boldsymbol{p}_q$ point resides. Once the cell is found, the closest point is determined.

If the dimension is larger than 2, the complexity of the Voronoi diagram increases to $\mathcal{O}(n^{\lceil k/2 \rceil})$.

TINN. Greenspan et al. give a method for determining closest points based on the triangle inequality (Triangle Inequality Nearest Neighbour) [50].

Their algorithm uses a reference point $\boldsymbol{r}$. In a pre-processing step the distances of all points $\boldsymbol{p} \in M$ to this reference point $\boldsymbol{r}$ are computed and stored together with $\boldsymbol{p}$ in a sorted list. During the point search the distance of the reference point $\boldsymbol{r}$ to the query point $\boldsymbol{p}_q$ is computed. Then a binary search is conducted to find candidate points. Afterwards, the list is travesed into both directions and all points $\boldsymbol{p}_i$ are tested until triangle inequality

$$|R_q - R_i| \leq D_{qi} \leq R_q + R_i$$

shows a minimum for D_{qi}. Here R_q is the Euclidean distance of the query point $\boldsymbol{p}_q$ to the reference point $\boldsymbol{r}$, i.e., $R_q = ||\boldsymbol{p}_q - \boldsymbol{r}||$ [50].
Even with employing binary search the run time of this algorithm is worse than searching using multi-dimensional trees.

The Cell Method. In contrast to the previous methods that optimize the search for the worst case, the following methods try to optimize the average search time. They make assumptions about the probability distribution of the input point set M. A simple method for solving the closest point problem is the cell method, also called Elias method.

The k-dimensional space is divided into small cells or voxels of equal size, i.e., a grid is computed. The axes of this grid are aligned with the axes of the coordinate system. The data in every cell is stored in a liked list. A spiral search in the cells starting from the query point finds always the closest point [40, 50, 105].

The expected value for the performance of this method depends on the assumed point distribution. If this distribution is uniform, the cell method is optimal [105]. The search time depends essentially on the number of points stored in a single cell. The spiral search can only be terminated, iff all cells whose distance to the query point is smaller or equal to the found minimum have been evaluated.

The memory consumption of this method is in the order of $\mathcal{O}(d^k)$ where d is proportional to the spatial distribution of the points. Therefore, this method is seldom used in practical robotics application.

The following subsections present methods for determining closest points that are based on multi-dimensional binary trees. Also approximation plays a role, i.e., the closest point is not computed exactly, but computed approximately.

k-d Trees for Solving the Closest Points Problem

k-d trees are a generalisation of binary search trees. Every node of the k-d tree represents a partition of the point set into two distinct sets, the successor nodes. The root of the tree represents the whole point set. The leaves of the tree are called buckets and are a partition of the set into small, disjunctive point sets. Furthermore, every node of the tree consists of the center and the dimension of the point set. Source code of an efficient implementation is given in [85].

Constructing k-d Trees. In k dimensions the entries have k keys. Every key might be used at the inner nodes to divide the point set. In the original k-d tree paper, these so-called split dimensions D have been chosen depending of the depth of node: $D = L \mod k + 1$ where L is the current depth of the tree. For the root is $L = 0$. Afterwards Bentley determines the split value randomly from the point set. The split dimension and value define an axis-aligned hyperplane in the k-dimensional space. The data are partitioned according to their position to the hyperplane into the successor nodes. The k-d tree is constructed until the number of points in the nodes falls below a threshold b (bucket size). Only the leaves of the tree contain the data points. Figure 4.3 visualizes the construction of the k-d tree.

For 3D data points we have $k = 3$. The 3-d tree contains separating planes that are parallel to the axes of the coordinate system, i.e., parallel to the planes $(x, y, 0)$, $(0, y, z)$ or $(x, 0, z)$.

Searching k-d Trees. Searching in k-d trees is done recursively. A given 3D point $\boldsymbol{p}_q$ needs to be compared to the separating plane (splitting dimension

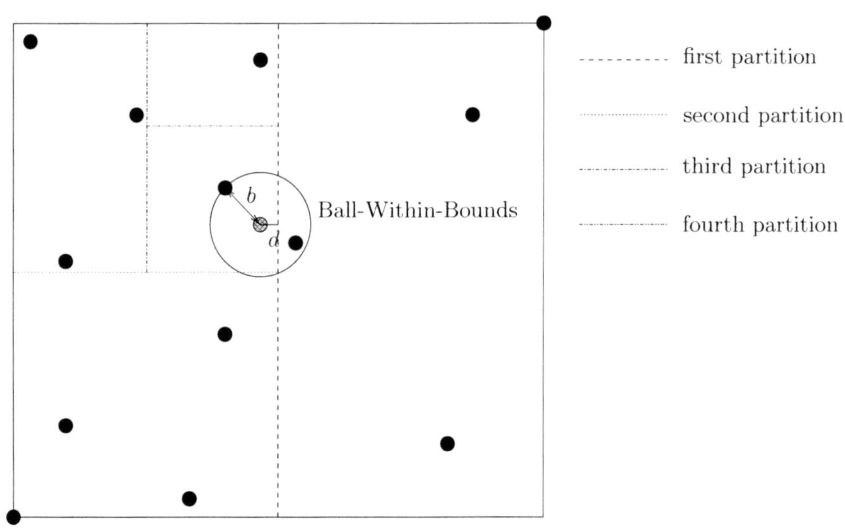

Fig. 4.3 Construction of a k-d tree (simplified). Here $k = 2$. After the quad that contains the whole point set is found it is partitioned into two smaller sets (first partition). Recursively additional hyperplanes, (here lines) are added. Searching k-d trees requires the computation of the distance b of query point $\boldsymbol{p}_q$ to the closest point in the bucket and a comparison to the distance to the closest border d. If the Ball-Within-Bounds test fails backtracking has to be implemented.

and splitting value) in order to decide on which side the search must continue. This procedure is executed until the leaves are reached. There, the algorithm has to evaluate all bucket points. However, the closest point may be in a different bucket, iff the distance d of the query point $\boldsymbol{p}_q$ to the limits is smaller than the one to the closest point in the bucket $\boldsymbol{p}_b$, i.e., iff

$$||\boldsymbol{p}_q - \boldsymbol{p}_b|| < d.$$

In this case backtracking has to be performed. Figure 4.3 shows a backtracking case, where the algorithms has to go back to the root of the tree. The test is known as Ball-Within-Bounds test [9, 40, 49].

The Optimized k-D Tree

The objective of optimizing k-d trees is to reduce the expected number of visited leaves. Three parameters are adjustable, namely, the split dimension and value, i.e., direction and place of the split hyperplane as well as the maximum number of points b in the buckets.

The solution of the optimization depends usually on the point distribution in the k-d tree and the distribution of the query points. Since one typically has no information about the query points k-d tree algorithms take only the distribution of the given points into account. For all possible queries this works sufficiently, but it will not be optimal for a specific query [40]. In addition the split dimension and the split value in a certain node are only determined by the points in this node. This enables the recursive construction and avoids the overall optimization that is known to be NP-complete [65].

Using these restrictions the split dimension and the split value have to be computed that enable the search algorithm to assign the given points to the successor nodes. Both nodes should occur with the same probability. This

Fig. 4.4 The optimized k-d tree uses splits along the longest axis to ensure compact volumes

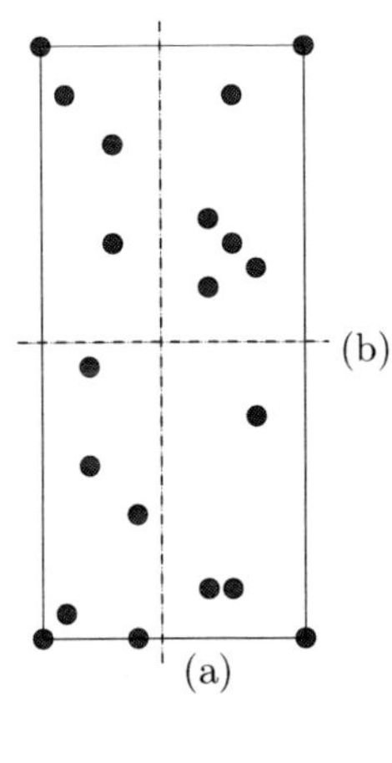

criterion implies that the partition has to be done along the median of the point set.

The search algorithm can prune its search, if the Ball-within-Bounds test fails, i.e., it can exclude branches if the distance to the partitioning hyperplane is larger than to the closest point (cf. Figure 4.3). The probability of excluding branches is maximal, if the split axis is oriented *perpendicular* to the longest axis to minimize the amount of backtracking. An example is given in Figure 4.4: If the set is partitioned along the axis (a) backtracking occurs more often than in the case (b), since possible Ball-within-Bounds intersect more likely with the hyperplane.

Performance Analysis. The optimized k-d tree is a k-d tree where the split dimension has been found using information about the dimensions of area covered by the stored points. The complexity of the tree construction is estimated as follows: In all depth all keys must be examined. This is proportional to kN. The depth of the tree is $\log N$, therefore the time for constructing the tree is in the order of $\mathcal{O}(kN \log N)$, which is the result of solving the recursion formula $T_N = 2T_{N/2} + kN$.

The expected value for the search time is determined in a geometric context. Assume $\boldsymbol{p}_i = (p_i(1), p_i(2), \ldots, p_i(k))$ are the keys for the ith data point in the k dimensional search space. All stored points are described by these values as well as the query point $\boldsymbol{p}_q$. The performance of the optimized k-d tree depends on the number of examined leaves that in turn could depend on the total number of stored points N, the dimensionality k and the number of points in the buckets b. Friedmann et al. prove that the last two parameters are not important [40].

Theorem 7. *The time for the search for the m closest points in an optimized k-d tree is logarithmic in the number of points in the tree N [40].*

Proof. Assume $S_m(\boldsymbol{p}_q)$ to be the smallest region in the coordinate system with center at $\boldsymbol{p}_q$, that contains the m closest points to the query $\boldsymbol{p}_q$:

$$S_m(\boldsymbol{p}_q) \equiv \{\boldsymbol{p} \mid ||\boldsymbol{p} - \boldsymbol{p}_q|| \leq ||\boldsymbol{p}_q - \boldsymbol{p}_m||\},$$

where $\boldsymbol{p}_m$ is the mth clostest point of $\boldsymbol{p}_q$ [40]. The k-dimensional volume $v_m(\boldsymbol{p}_q)$ is given by

$$v_m(\boldsymbol{p}_q) = \int\limits_{S_m(\boldsymbol{p}_q)} d\boldsymbol{p}.$$

The probalility $u_m(\boldsymbol{p}_q)$ for a point in this region is defined as

$$u_m(\boldsymbol{p}_q) \equiv \int\limits_{S_m(\boldsymbol{p}_q)} p(\boldsymbol{p})\, d\boldsymbol{p}, \qquad \text{where} \quad 0 \leq u_m(\boldsymbol{p}_q) \leq 1. \tag{4.39}$$

It can be shown that the probability distribution $u_m(\boldsymbol{p})$ is a beta-distribution $B(m, N)$ [40, 44] which means that

$$p(u_m) = \frac{N}{(m-1)!(N-m)!}(u_m)^{m-1}(1-u_m)^{N-m}$$

is independent of the probability density of the points $p(\boldsymbol{p})$ and the used norm. The expected value of this distribution is calculated as

$$E[u_m] = \int_0^1 u_m \; p(u_m) \; du_m = \frac{m}{N+1}. \tag{4.40}$$

This result means that every compact volume that contains m points has at average the probability of $m/(N+1)$.

For the following considerations lets suppose N to be large enough and $S_m(\boldsymbol{p}_q)$ to be small. Therefore, we can assume that the probability distribution $p(\boldsymbol{p})$ is nearly constant within the region $S_m(\boldsymbol{p}_q)$ [40]. The Eq. (4.39) is approximated by

$$u_m(\boldsymbol{p}_q) \approx \bar{p}(\boldsymbol{p}_q) v_m(\boldsymbol{p}_q)$$

and Eq. (4.40) by

$$E[v_m(\boldsymbol{p}_q)] \approx \frac{m}{N+1} \frac{1}{\bar{p}(\boldsymbol{p}_q)}. \tag{4.41}$$

$\bar{p}(\boldsymbol{p}_q)$ is the constant probability density over the small region $S_m(\boldsymbol{p}_q)$, that is always larger than zero. Let's consider now the way the dK-d tree is constructed. Since the algorithm uses the median as split value we assume that every leaf contains roughly the same number of data points b. The choice of the split dimension ensures that the data points in the leaves are compact (cf. Figure 4.4). The expected geometric form of the leaves is a hypercubus where edge length is the kth square root of the leaf volume. Furthermore, the hypercubus is axes aligned. From Eq. (4.41) the expected volume is calculated as:

$$E[v_b(\boldsymbol{p}_b)] \approx \frac{b}{N+1} \frac{1}{\bar{p}(\boldsymbol{p}_b)}, \tag{4.42}$$

where $\boldsymbol{p}_b$ is the point that localizes the leaf in the coordinate space.

Given the smallest hypercubus that contains the region $S_m(\boldsymbol{p}_q)$ and whose edges are parallel with the coordinate system axes. The volume of the hypercubus $V_m(\boldsymbol{p}_q)$ is proportional to $v_m(\boldsymbol{p}_q)$. The constant of proportionality $g(k)$ depends on the dimensionality k [40]. It holds:

$$V_m(\boldsymbol{p}_q) = g(k)\ v_m(\boldsymbol{p}_q) \qquad \text{and} \tag{4.43}$$

$$E[V_m(\boldsymbol{p}_q)] = \frac{m}{N+1}\frac{g(k)}{\bar{p}(\boldsymbol{p}_q)}. \tag{4.44}$$

To estimate the expected number of leaves that have to be analysed by the search algorithm one has to compute the average number of leaves $\bar{l}$ that overlap with the region $S_m(\boldsymbol{p}_q)$. An upper bound for this number is given by the average number of leaves $\bar{L}$ that enclose the hypercubus $S_m(\boldsymbol{p}_q)$. $\bar{L}$ is given by

$$\bar{L} = \left(\frac{e_m(\boldsymbol{p}_q)}{e_b(\boldsymbol{p}_q)+1}\right)^k. \tag{4.45}$$

where $e_m(\boldsymbol{p}_q)$ is the edge length of the hyperkubus, that contains $S_m(\boldsymbol{p}_q)$ and $e_b(\boldsymbol{p}_q)$ is the edge length of the adjacent hypercubus leaves. The edge length of the hypercubus is calculated uing the kth root of the volume. If we consider the Eq. (4.42) to Eq. (4.45) we derive

$$\bar{l} \leq \bar{L} = \left(\left(\frac{m}{b}g(k)\right)^{1/k}+1\right)^k$$

as the upper bound for the average number of leaves that overlap with the constant volume $S_m(\boldsymbol{p}_q)$. Since the number of data points in every leaf is b, we have for the upper bound $\bar{R}$ for the average number of data that have to be analyzed:

$$\bar{R} \leq b\bar{L} = b\left(\left(\frac{m}{b}g(k)\right)^{1/k}+1\right)^k. \tag{4.46}$$

Eq. (4.46) gives two results [40]: First, if we perform a minimization of $\bar{R}$ using b than we have $b = 1$ as an optimal value. This means that leaves should contain exactly one data point to search the optimized k-d tree as fast as possible. However, an analyses of implementations show that the optimal value is slightly larger, and lies between 10 to 100 points (cf. Figure 4.5). Second, the expected number of leaves that have to be analyzed is independent of the overall number of points N that are stored in the k-d tree and independent of the distribution of keys $p(\boldsymbol{p})$.

These results can be understood intuitively [40]: The k-d tree algorithm aims at minimizing the overlapping area to a volume. Thus, the partitions must be as small as possible. That there is no dependence between the number of leaves that have to be analyzed and the number of point N or the distribution of keys is a direct consequence of the tree construction algorithm for the optimized tree. The partitioning is done such that every leaf has the same probability to hold the region $S_m(\boldsymbol{p}_q)$ with the m closest points. The partitioning is done such that the geometric structure becomes compact and every leaf contains the same

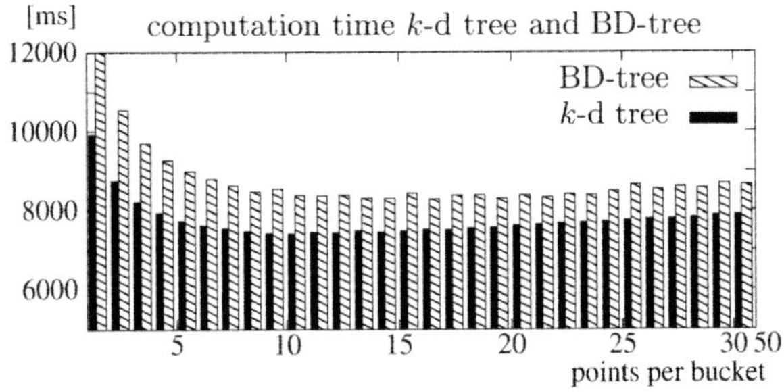

Fig. 4.5 Computation time for closest point computation using k-d trees and BD-trees (accumulated values) with different maximal numbers of points per leaf

number of data points. The dependence of the volume of the leaves on the number of data point and distribution of the keys is identical to the one of the region $S_m(\boldsymbol{p}_q)$, that contains the m closest points. If the total number of data or the local key density are increased then the volumes of a leaf regions is lowered by the same rate. Therefore the number of overlapping leaves $\bar{l}$ remains constant.

Since a constant number of leaves has to be searched, closest points are found in logarithmic time. The k-d tree is a balanced binary tree. The time needed to reach the leaves from the root depends logarithmically on the number of nodes that in turn is proportional to the number of data points N. The cost of backtracking is in average proportional to $\bar{l}$ and independent of N. Therefore m closest points or the closest point are found in logarithmic time. □

Computing the Median

The performance analysis of the optimized k-d tree works, since the point clouds are splitted at the median. Thus it is ensured, that the points are equally distributed to the successor nodes. Computing the median must be done if linear time, to enable the construction of a k-d tree in $\mathcal{O}(N \log N)$ time. Amazingly the following simple algorithm solves the problem in a randomized fashion:

Algorithm 4.2. Randomized Computation of the Median. The ith smallest element of the input array `A[p,...,r]` is returned.

```
Randomized-Select(A,p,r,i)
if (p == r) return A[p];
q = Randomized-Partition(A,p,r)
k = q - p + 1;
if (i <= k) return Randomized-Select(A,p,q,i)
      else  return Randomized-Select(A,q+1,r,i-k)
```

The function `Randomized-Partition(A,p,r)` splits the array `A` into two non-empty subsets, such that all elements in `A[p,...,q]` are smaller than all elements in `A[q,...,r]`. A runtime analysis shows this algorithm is in $\Theta(n^2)$ in worst case. However, in the average case we need only $\Theta(n)$ time [21].

In [21] there is a deterministic $\Theta(n)$ algorithm for computing the median given. The algorithm uses a direct (deterministic) method to compute a "good" pivot element for the partition. Unfortunately, this method shows large constants and therefore should only improve the computation of the median in point clouds with unusual point distributions.

Approximate Calculation of Closest Points

Greenspan and Yurik have presented in 2003 the approximated k-d tree for the ICP algorithm [49]. The construction of an approximated k-d tree equals the one of the optimized k-d tree. The idea of the approximation is to use the closest point in the bucket $\boldsymbol{p}_b$ where the query point $\boldsymbol{p}_q$ resides as closest point for the ICP algorithm. Thus $\boldsymbol{p}_b$ is determined by depth-first search only, and the Ball-Within-Bounds test as well as backtracking are avoided [49,90]. Greenspan and Yurik have showed that the ICP algorithm stops in a local minimum that is close to the one that is obtained in the non-approximated case.

This approximated k-d tree is constructed in $\mathcal{O}(N \log N)$ time and the search time is now in $\Theta(\log N)$. In addition to the described approximation one can use the mean points in the bucket as approximated closest point avoiding the search over all leaf elements. This increases the approximation and the run time is even lower [90].

Unfortunately, the approximation scheme described above does not allow to assure the quality of the closest point, i.e., it might be arbitrarily bad. Ayra et al. introduce 1993 the following notation to overcome this problem: Given $\varepsilon > 0$, the point $\boldsymbol{p} \in D$ is the $(1+\varepsilon)$-approximated nearest neighbour of $\boldsymbol{p}_q$, if

$$||\boldsymbol{p} - \boldsymbol{p}_q|| \leq (1+\varepsilon)\ ||\boldsymbol{p}^* - \boldsymbol{p}_q||\,,$$

where $\boldsymbol{p}^*$ is the true nearest neighbour (c.f. Figure 4.6). In other words: $\boldsymbol{p}$ is within a relative error ε from the true nearest neighbor.

To apply this notation, the search program of the k-d tree has to record at any time the closest point $\boldsymbol{p}$ found so far. Searching is terminated and $\boldsymbol{p}$ is returned, if the distance to the leaves that have to be analyzed exceeds the value of

$$||\boldsymbol{p}_q - \boldsymbol{p}||\,/(1+\varepsilon)$$

Figure 4.15 shows the run time in a search tree depending on the value of ε and the number of data points per leaf.

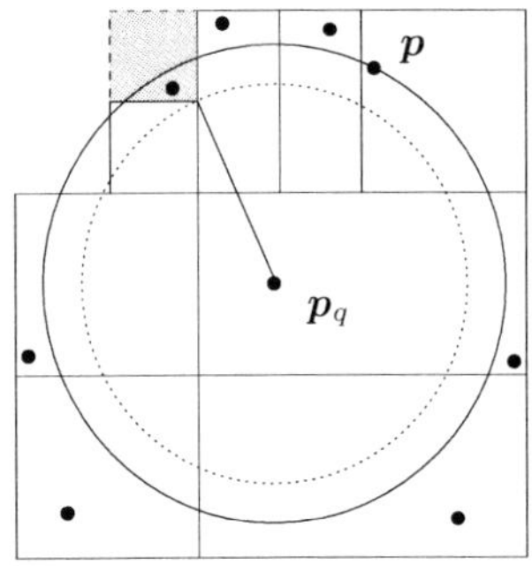

Fig. 4.6 The $(1+\varepsilon)$-approximated nearest neighbor. The grey cell does not have to be analyzed, since $\boldsymbol{p}$ fulfills already the approximation critera. Based on: [4].

Balanced Box-Decomposion Trees for Computing Closest Points

Arya et al. [4] introduced in 1998 an optimal algorithm for computing approximate nearest neighbours for fixed dimensions. Their algorithm uses balanced box-decomposion trees (BD-trees) as primary data structure. The BD-tree combines two important properties of geometrical data structures. First, the *cardinality* of points that lie on path though the tree is reduced exponentially. This is similar to the optimized k-d tree. Second, the *aspect ratio* of the longest and smallest side of the hypercubus is bounded by a constant. For example a quadtree is a data structure that also shows this property. This assurance cannot be made by the optimized k-d tree.

Construction of a BD-tree. A BD-tree is constructed using the repetitive application of two operations: Splits and shrinks [4]. They divide the data set in two successor nodes, thus resulting in a binary tree. The split operation uses a hyperplane that is aligned with the coordinate system and the median as split key as in the optimized k-d tree case. The shrink operation determines two disjunctive subsets by computing a hyperbox that is subtracted from the

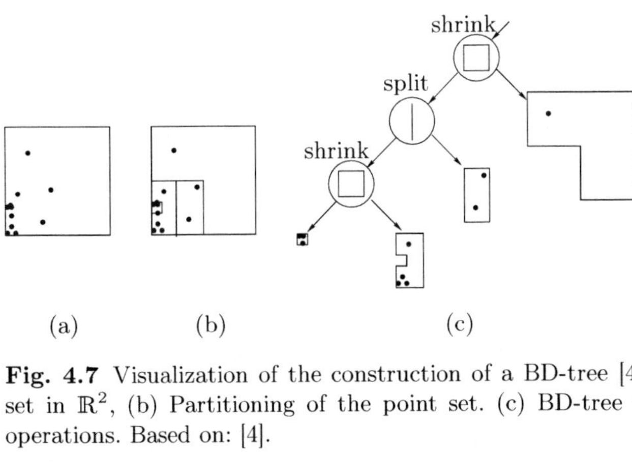

Fig. 4.7 Visualization of the construction of a BD-tree [4]. (a) The input point set in $\mathbb{R}^2$, (b) Partitioning of the point set. (c) BD-tree with shrinks and split operations. Based on: [4].

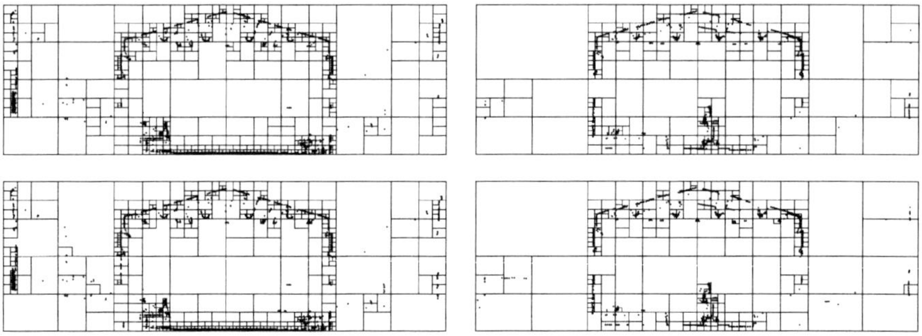

Fig. 4.8 Plots of a k-d tree (top) and a BD-tree (bottom) of the point set given in Figure 4.1. Since the dimension is $k = 3$ (x, y)-projections have been extracted with z=100 cm (left) and z=500 cm.

node. Here the set operation subtraction is used. Figure 4.7 sketches the construction of a BD-tree.

Figure 4.8 shows a k-d tree and a BD-tree for the point set given in Figure 4.1.

Searching in a BD-tree. The search in a BD-tree is similar to the one used for the optimized k-d tree, i.e., first of all the leaf is determined that contains the query point $\boldsymbol{p}_q$. Afterwards the Ball-Within-Bounds test is applied and backtracking is started until all possible leaves have been searched. An approximate computation of the closest points is also straightforward.

For the search algorithm in balanced BD-trees Ayra et al. prove the optimality, i.e., the following theorem [4]:

Theorem 8. *Assume S to be a set of n data points in $\mathbb{R}^k$. There exists a constant $c_{k,\varepsilon} \leq k\lceil 1 + 6k/\varepsilon \rceil^k$ such that it is possible to construct a data structure in $\mathcal{O}(kn \log n)$ time, namely a BD-tree and that fulfils the following condition: For a given $\varepsilon > 0$ and $\boldsymbol{q} \in \mathbb{R}^k$ an $(1 + \varepsilon)$-approximated nearest neighbour of $\boldsymbol{p}_q$ is found in $\mathcal{O}(c_{k,\varepsilon} \log n)$ time.*

Proof. See [4]. □

Searching a balanced BD-tree requires less operations than searching a k-d tree necessary, since through the shrink operations it should be possible to focus the search to areas with high point densities. However an analysis shows that the k-d tree slightly outperforms the BD-tree in the dimension $k = 3$ (cf. Figure 4.5).

Cached k-d Tree Search

k-d trees with caching contain, in addition to the limits of the represented point set and to the two child node pointers, one additional pointer to the predecessor node. The root node contains a null pointer. During the recursive

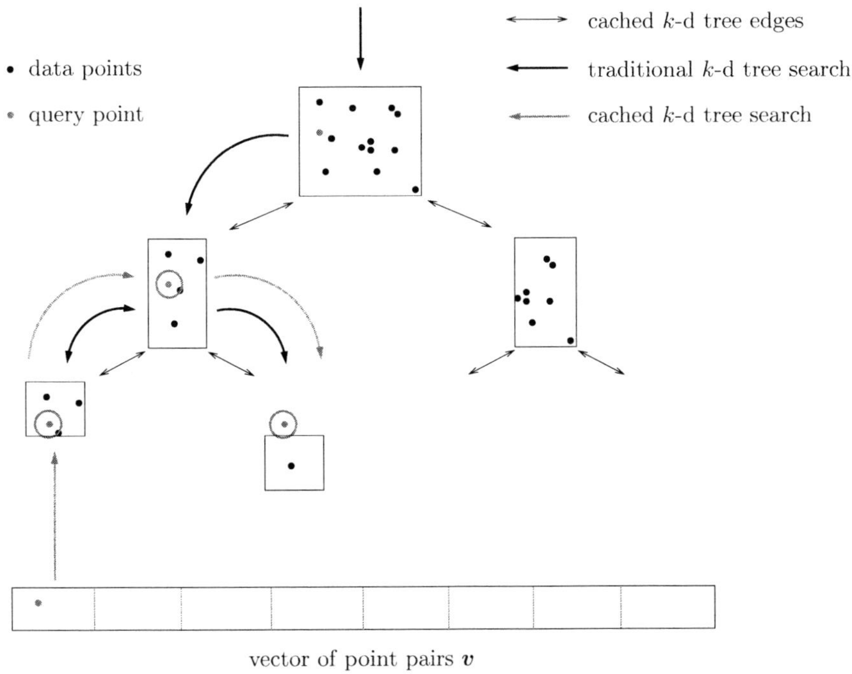

Fig. 4.9 Schematic description of the cached k-d tree search method: Instead of closest point searching from the root of the tree to the leaves that contain the data points, a pointer to the leaves is cached. In the second and following ICP iterations, the tree is searched backwards.

construction of the tree, this information is available and no extra computations are required.

For the ICP algorithm, lets distinguish between the first and the following iterations: In the first iteration, a normal k-d tree search is used to compute the closest points. However, the return function of the tree is altered, such that in addition to the closest point, the pointer to the leaf containing the closest point is returned and stored in the vector of point pairs. This supplementary information forms the cache for future look-ups.

In the following iterations, these stored pointers are used to start the search. If the query point is located in the bucket, the bucket is searched and the ball-within-bounds test is applied. Backtracking is started, iff the ball lies not completely within the bucket. If the query point is not located within the bucket, then backtracking is started, too. Since the search is started in the leaf node, explicit backtracking through the tree has to be implemented using the pointers to the predecessor nodes (see Figure 4.9). Algorithm 4.3 summarizes the ICP with cached k-d tree search.

Algorithm 4.3. ICP with cached k-d tree search

for $i = 0$ to *maxIterations* **do**
  if $i == 0$ **then**
    for all $\boldsymbol{d}_j \in D$ **do**
      search k-d tree of set M top down for point $\boldsymbol{d}_j$
      $\boldsymbol{v}_j = \left(\boldsymbol{d}_j, \boldsymbol{m}_{f(\boldsymbol{d}_j)}, \text{ptr\_to\_bucket}(\boldsymbol{m}_{f(\boldsymbol{d}_j)})\right)$
    end for
  else
    for all $\boldsymbol{d}_j \in D$ **do**
      search k-d tree of set M bottom up for point $\boldsymbol{d}_j$ using ptr_to_bucket($\boldsymbol{m}_{f(\boldsymbol{d}_j)}$)
      $\boldsymbol{v}_j = (\boldsymbol{d}_j, \boldsymbol{m}_{f(\boldsymbol{d}_j)}, \text{ptr\_to\_bucket } \boldsymbol{m}_{f(\boldsymbol{d}_j)})$
    end for
  end if
  calculate transformation $(\boldsymbol{R}, \boldsymbol{t})$ that minimizes the error function Eq. (4.2)
  apply transformation to data set D
end for

Performance of Cached k-d Tree Search. The proposed cached k-d tree search needs $\mathcal{O}((I + \log N_m)N_d)$ time in the best case. This performance is reached if constant time is needed for backtracking, resulting in $N_d \log N_m$ time needed for constructing the tree, and $I \cdot N_d$ for searching in case no backtracking is necessary. Obviously the backtracking time depends on the computed ICP transformation $(\boldsymbol{R}, \boldsymbol{t})$. For small transformations the time is nearly constant.

Cached k-d tree search needs $\mathcal{O}(N_d)$ extra memory for the vector $\boldsymbol{v}$, i.e., for storing the pointers to the tree leaves. Furthermore, additional $\mathcal{O}(N_m)$ memory is needed for storing the backwards pointers in the k-d tree.

4.1.5 Implementation Issues

The first step of the ICP algorithm, the calculation of the point pairs is computationally expensive. the naïve implementation needs $\mathcal{O}(n^2)$ which can be reduced to the order of $\mathcal{O}(n \log n)$. In contrast, the calculation of the optimal transformation is in $\mathcal{O}(n)$. Applying the optimal transformation to the points in the data set needs again only $\mathcal{O}(n)$ time. Thus, the overall run time of ICP is dominated by the calculation of closest points.

Data reduction reduces the time needed for scan matching, i.e., for computing closest points. Different methods, e.g., random or uniform point selection are well suited for the reduction. Sophisticated strategies incorporate the point normals or use the so called covariance sampling [45, 107]. The first method selects points such that the surface normals face into all directions. Covariance sampling selects uniformly the points and afterwards the point pairs that lead to unstable transformations are neglected by estimating a covariance matrix. This avoids the sliding of two featureless areas in 3D scans [45].

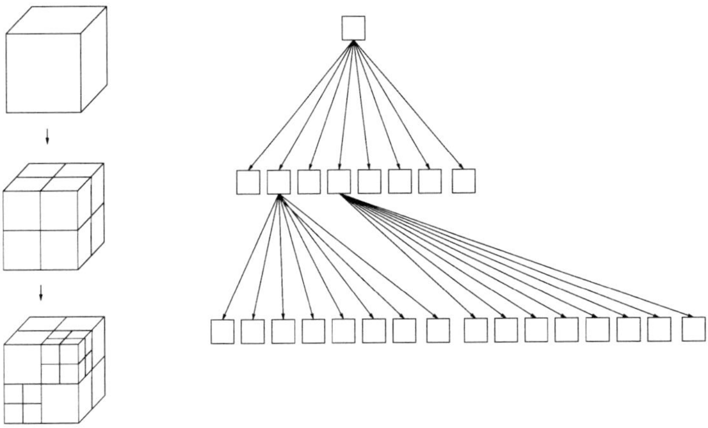

Fig. 4.10 An octree represents a cube, that is splitted into 8 successor nodes. Left: Recursive subdivision. Right: The corresponding octree.

Boulanger et al. [14] postulate that subsampling 3D range images should maintain the manifold of the scanned surface, i.e., the structure of the scene has to be represented by the subsampled points as it is in the original case. The following simple method is the de facto standard for subsampling 3D point cloud data, namely octree based subsampling. Starting with a cube containing the whole 3D point cloud, the cube is divided into 8 smaller, equally-sized cubes by 3 planes (see Figure 4.10). After every split it is checked whether the cube is empty or not. Empty cubes are pruned and the subdivision stops if a certain discretization level is reached. Finally, the center of the cubes are the subsampled set of points.

Figure 4.12 visualizes the octree based subsampling. Obviously, the maximum number of octants is bounded by the number of 3D data points. Computing too many subdivisions yields octants around every point and no reduction is performed, thus the minimal octant size is the important threshold while subsampling.

Fig. 4.11 3D point cloud

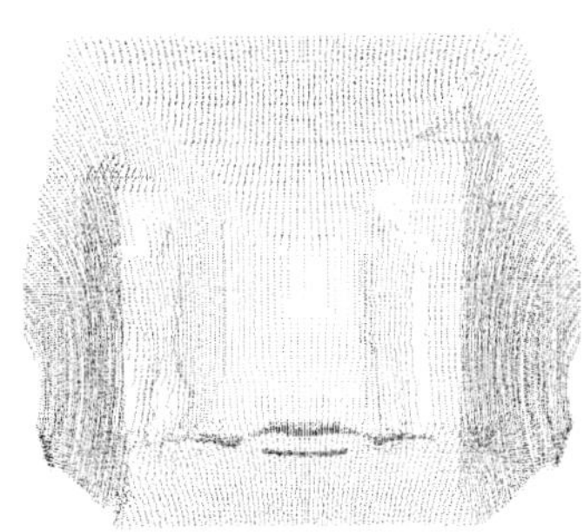

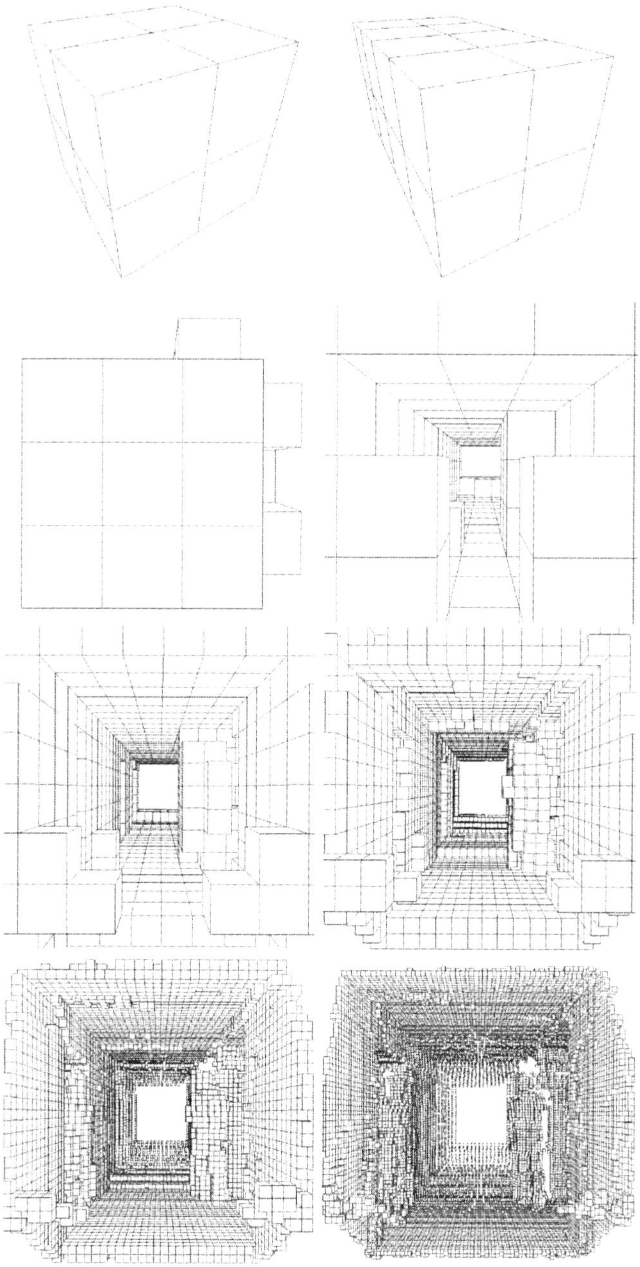

Fig. 4.12 Octree construction for the 3D point cloud given in Figure 4.11. Subsampling is perfomed by taking the centroids of the octants at a certain depth of the tree.

4.1.6 The Parallel ICP Algorithm

The basic work for parallelization of the ICP algorithm (pICP) was done by Langis et al. [72]. pICP showed compelling results on a computer cluster with p processors for registration of large data sets. The basic idea is to divide the data set D into p parts and to send these parts together with the whole model set M to the child processes that compute concurrently the expensive closest point queries. Afterwards, these point correspondences are transmitted back to the parent, that uses them for computing the transformation by minimizing Eq. (4.2). Then this transformation is sent to the childs, which transform the data set D. The process is iterated until the minimum of Eq. (4.2) is reached.

The above parallelization scheme for ICP uses a lot of bandwidth for transmitting corresponding points. Thus, Langis et al. [72] proposed several enhancements to the parallel method. One of these improvements is avoiding the transfer of the corresponding points. The computation of the correlation matrix $\boldsymbol{H}$ in Eq. (4.9) is parallelized and partially executed by the child processes, i.e., Eq. (4.9) is transformed as follows

$$\begin{aligned} \boldsymbol{H} &= \sum_{i=1}^{N} (\boldsymbol{m}_i - \boldsymbol{c}_m)(\boldsymbol{d}_i - \boldsymbol{c}_d)^T \\ &= \sum_{i=1}^{p} \sum_{j=1}^{N_i} (\boldsymbol{m}_{i,j} - \boldsymbol{c}_m)(\boldsymbol{d}_{i,j} - \boldsymbol{c}_d)^T, \end{aligned} \tag{4.47}$$

where N_i denotes the number of points of child i, $\boldsymbol{c}_{m_i}, \boldsymbol{c}_{d_i}$ the centroid of points of child i, and $(\boldsymbol{m}_{i,j}, \boldsymbol{d}_{i,j})$ the jth corresponding point of child i. Furthermore, for Eq. (4.47) holds

$$\begin{aligned} \boldsymbol{H} &= \sum_{i=1}^{p} \sum_{j=1}^{N_i} \Big((\boldsymbol{m}_{i,j} - \boldsymbol{c}_{m_i} + (\boldsymbol{c}_{m_i} - \boldsymbol{c}_m)) \\ &\qquad (\boldsymbol{d}_{i,j} - \boldsymbol{c}_{d_i} + (\boldsymbol{c}_{d_i} - \boldsymbol{c}_d))^T \Big) \\ &= \sum_{i=1}^{p} \left(\boldsymbol{H}_i + N_i (\boldsymbol{c}_{m_i} - \boldsymbol{c}_m)(\boldsymbol{c}_{d_i} - \boldsymbol{c}_d)^T \right) \end{aligned} \tag{4.48}$$

$$\text{where, } \boldsymbol{H}_i = \sum_{j=1}^{N_i} (\boldsymbol{m}_{i,j} - \boldsymbol{c}_{m_i})(\boldsymbol{d}_{i,j} - \boldsymbol{c}_{d_i})^T \tag{4.49}$$

Therefore, after the point correspondences are parallel computed. Eq. (4.48) and (4.49) form the algorithm for computing the correlation matrix $\boldsymbol{H}$ in a parallel fashion. The centroids $\boldsymbol{c}_m$ and $\boldsymbol{c}_d$ (cf. Eq. (4.3)) are also computed from these intermediate results, i.e.,

$$\boldsymbol{c}_m = \frac{1}{N}\sum_{i=1}^{p} N_i \boldsymbol{c}_{m_i}, \qquad \boldsymbol{c}_d = \frac{1}{N}\sum_{i=1}^{p} N_i \boldsymbol{c}_{d_j}$$

Finally, after the parent has collected the vectors $\boldsymbol{c}_{m_i}$, $\boldsymbol{c}_{d_i}$ and the matrix $\boldsymbol{H}_i$ from the childs, it is possible to compute the transformation $(\boldsymbol{R}, \boldsymbol{t})$ to align the point sets.

Parallelization of *k*-d tree Search

Here we consider an OpenMP implementation of the pICP algorithm, that is a shared memory architecture with p processors or threads (symmetric multi-processing). Therefore, there is no need to create p copies of the model set M. However the search is executed in parallel using parallel k-d tree search.

Parallel Construction of *k*-d trees. Since the k-d tree partitions the point set into two disjunctive subsets, the construction is easily partitioned by executing the two recursive cases in parallel.

Parallel Search of *k*-d trees. In the pICP algorithm several search requests have to be handled by the k-d tree at the same time. Efficient implementations of k-d tree search algorithms use static, i.e., global variables to save the current closest point. This current point is first derived by depth first search and then updated in the backtracking phase. When executing the search with parallelism p copies of this point must be created.

Implementation Details. Most high performance processors insert a hardware cache buffer between slow memory and the high speed registers of the CPU. If several threads use unique data elements in the same hardware cache line for read and write, then so-called false sharing might occur. If one of the threads writes to a cache line, the same cache line referenced by the second thread is invalidated of the cache. Any new references to data in this cache line by the second thread results in a cache miss and the data has to be loaded again from memory. To avoid false sharings, we pad each thread's data element to ensure that elements owned by different threads all lie on separate cache lines as follows:

```
class KDParams {
public:
  /** pointer to the closest point.
    *  size = 4 bytes of 32 bit machines
    */
  double *closest;
  /** distance to the closest point.
    * size = 8 bytes
    */
```

```
    double closest_d2;
    /** pointer to the point,
      * size = 4 bytes of 32 bit machines
      */
    double *p;
    /** expand to 128 bytes to avoid
      * false-sharing, 16 bytes from above
      * + 28*4 bytes = 128 bytes */
    int padding[28];
  };
```

In our search tree these padded parameters are included using memory alignment:

```
  class KDtree {
    // [snip]
  public:
     __declspec (align(16)) \
        static KDParams params[MAX_OPENMP_NUM_THREADS];
  };
```

Parallel Search of Cached k-d trees. Searching cached k-d trees in a parallel fashion is accomplished by placing the `for`-statements of Algorithm 4.3 in OpenMP directives. The cache is currently a section of the main memory, thus all threads are allowed to access the cached data via shared memory.

4.2 Evaluation of the ICP Algorithm

Point Correspondences. The first step of the ICP algorithm computes the point correspondences $w_{i,j}$. Here a threshold for the maximal point-to-point distance is used. If these distances are smaller than $d_{\max}$ the point pair is created. Setting $d_{\max} = \infty$ we compute for all points in set D the closest point in set M and use it for calculating resulting transformation. In this case Besl and McKay have shown that the value $E(\boldsymbol{R}, \boldsymbol{t})$ is decreasing in a monotone fashion and since the value is bounded the ICP algorithm converges to a local minimum. However, using all of these point pairs is at a disadvantage for partially overlapping data sets and usually lowers the quality of the scan matching. Partially overlapping data sets contain many data points which do not describe the same point in space and therefore should not have correspondences. The assumption is, that using $d_{\max}$ and a good starting guess let the ICP algorithm converge to the correct minimum.

Using a suited threshold for $d_{\max}$ we achieve that in every iteration of the ICP algorithm point pairs are generated that describe the same point in space. Figure 4.13 shows the dependence of the quality of the scan matching on the threshold $d_{\max}$. The ground truth pose was determied by an independent measurement system and the quality is calculated using 3 DoF only, i.e., by

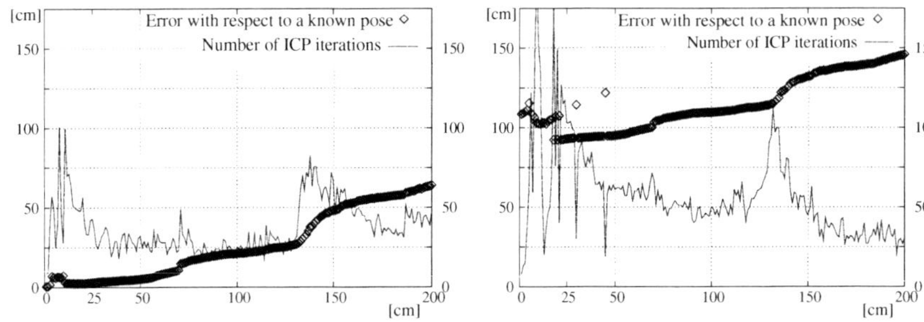

Fig. 4.13 Quality of the scan matching depending on the threshold $d_{\max}$ (Error function (4.50) and number of ICP iterations until convergence. Left: Starting pose is set to the ground truth pose. Right: Shifted and rotated start pose.

$$e(x, z, \theta) = ||(x, z) - (x_{\text{ref}}, z_{\text{ref}})|| + \gamma \, ||\theta - \theta_{\text{ref}}|| \, . \tag{4.50}$$

The following criterions for choosing $d_{\max}$ have to be considered:

1. The resolution of the 3D scans: The distance between points of the scan influence the parameter $d_{\max}$. If the environment is sampled with a coarse resolution, $d_{\max}$ should be large. This implies the threshold depends on the used hardware.
2. The initial pose of the 3D scan play an important role. If the start pose is off the correct pose, choosing a large $d_{\max}$ might help that the ICP algorithm converges to the correct minimum. With a large $d_{\max}$ ICP uses more correspondences.
3. To yield the best possible scan matching, $d_{\max}$ should be small.

Based on these 3 conditions, the following two step strategy seems appropriate: Starting with a large threshold, e.g., $d_{\max} = 25\,\text{cm}$ for indoor environments sampled in a radial fashion with 1°, for a fixed number of iterations this threshold is adjusted in a second step to some lower value.

ICP offers the possibility to weight the point correspondences using $w_{i,j}$ (see Eq. (4.1)). Assigning the weights according to the following rules has been studied in [85]:

1. Constants weights:

$$w_{i,j} = \begin{cases} 1 & \text{, if } \boldsymbol{d}_j \text{ is the closest point to } \boldsymbol{m}_i \text{ and the point-to-point} \\ & \text{distance is smaller than } d_{\max}, \\ 0 & \text{, otherwise.} \end{cases}$$

2. Weight depending on the distance:

$$w_{i,j} = \begin{cases} 1 - \frac{||\boldsymbol{m}_i - \boldsymbol{d}_j||}{d_{\max}} & \text{, if } \boldsymbol{d}_j \text{ is the closest point to } \boldsymbol{m}_i \text{ and} \\ & ||\boldsymbol{m}_i - \boldsymbol{d}_j|| < d_{\max}, \\ 0 & \text{, otherwise.} \end{cases}$$

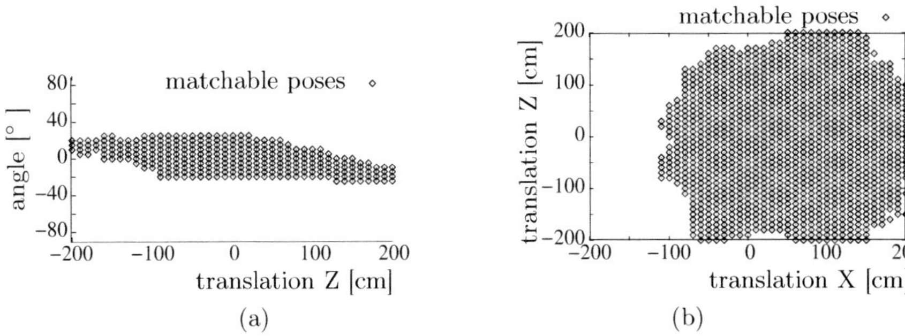

Fig. 4.14 Starting poses that lead to a correct matching

3. Weights depending on addional information, e.g., intensity data:

$$w_{i,j} = \begin{cases} 1 - |I(\boldsymbol{m}_i) - I(\boldsymbol{d}_j)| & \text{, if } \boldsymbol{d}_j \text{ is the closest point to } \boldsymbol{m}_i \text{ and} \\ & ||\boldsymbol{m}_i - \boldsymbol{d}_j|| < d_{\max}, \\ 0 & \text{, otherwise.} \end{cases}$$

However, it turned out that the possibility to weight point pairs has no major influence to ICP scan matching [85].

Robustness. To use scan matching on mobile robots we have to analyse which scans are matchable, depending on the quality of the starting guess. Such an analysis is quite difficult, since the genuine truth is difficult to obtain. Section 6.4 presents an automated benchmarking technique for large scale 6D SLAM. Figure 4.14 shows matchable start poses (3 DoF) for ICP scan matching evaluated by Eq. (4.50) (cf. also Figure 4.13).

In general, the following hands-on-experiences can be reported:

1. Rotational errors are harder to correct than translational ones. Therefore, the sensors of the robot must guess the rotation more precisely. This holds true for all Euler angles, since there is no distinction of them in the algorithm.
2. The geometry of the scanned scenes has an important impact on the result of scan matching. Unstructured scenes cannot be matched, for example two sampled planes slide and error minimization is not possible. Scenes that are highly structured, can be matched reliably. For example, the rotation estimation causes less problems in office corridors or tunnels in mines compared to spacious outdoor environments.

Efficiency of the Approximated k-d Tree and BBD-tree. Figures 4.5 and 4.15 present the computing times for scan matching using a k-d tree or BD-tree and for the approximated variants. The approximated variants show better performance, and with some exceptions k-d trees outperform BD-trees. Figure 4.16 presents the starting poses that lead to a correct scan matching

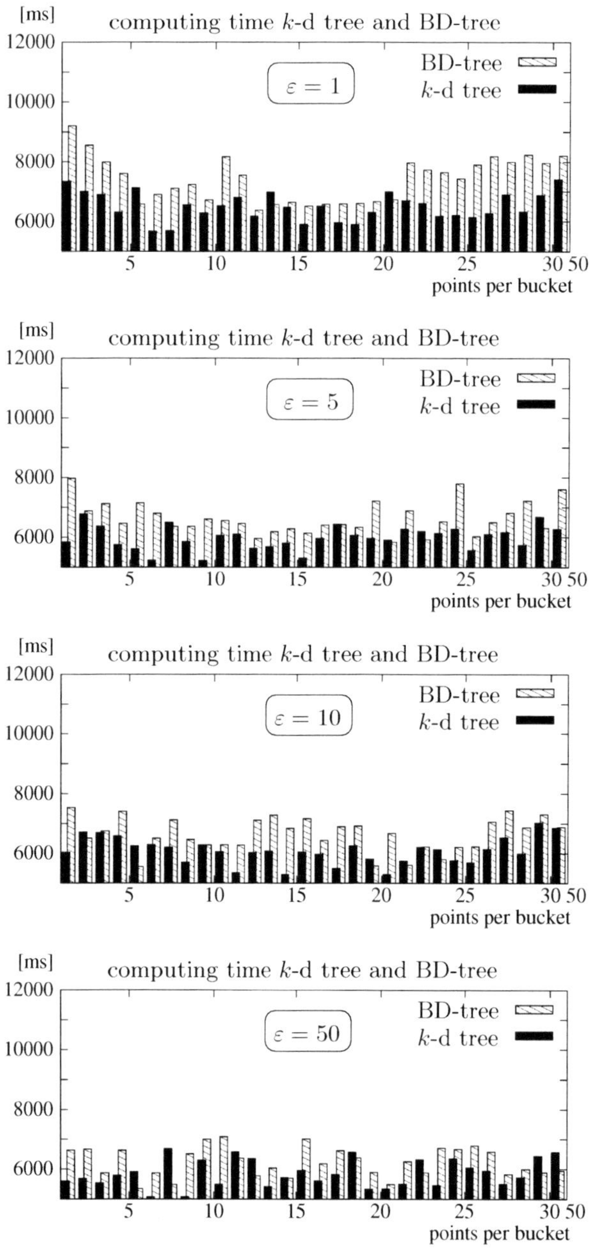

Fig. 4.15 Computing time for scan matching depending on the number of data points in the leaves and on approximation. Results for $\varepsilon = 1$, $\varepsilon = 5$, $\varepsilon = 10$, $\varepsilon = 50$ are given. Without approximation the times are given in Figure 4.5.

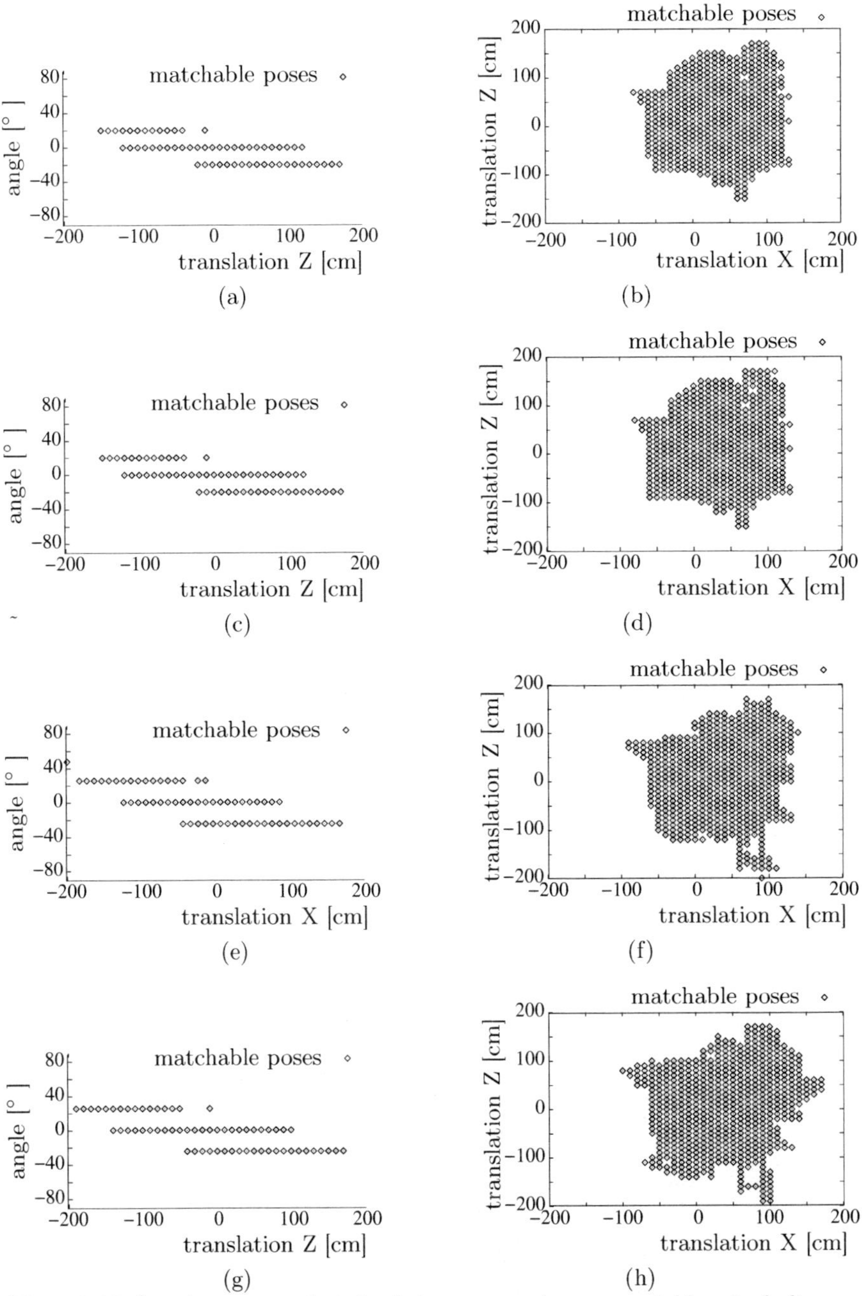

Fig. 4.16 Starting poses that lead to a correct scan matching including ε-approximated search for closest points. Different values for the approximation ε and b the number of data points in the leaf are presented: (a), (b): $\varepsilon = 1$ and $b = 10$. (c), (d): $\varepsilon = 1$ and $b = 20$. (e), (f): $\varepsilon = 10$ and $b = 10$. (g), (h): $\varepsilon = 50$ and $b = 5$.

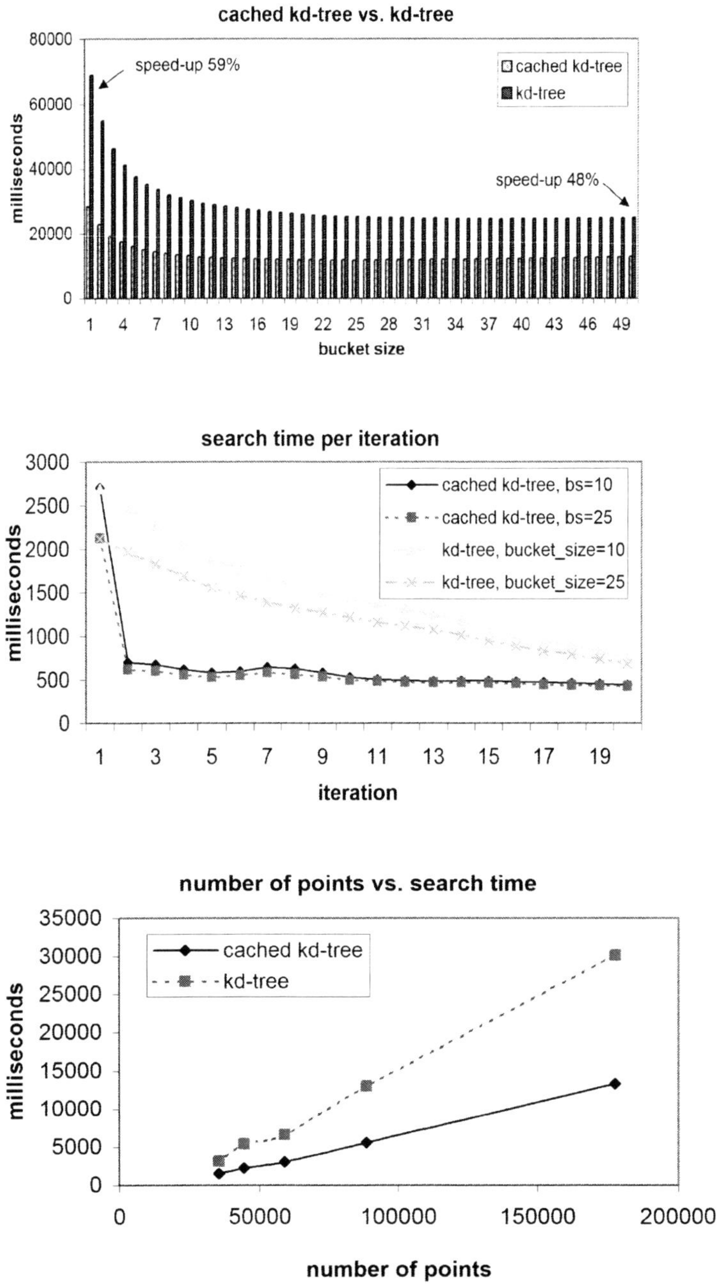

Fig. 4.17 Detailed results for a scan matching experiment. Top: Search time vs. bucket size. Middle: Search time per iteration for bucket sizes 10 and 25. Bottom: Search time depending on the number of points.

for different values of ε and b, i.e. for different granularity of approximation and number of 3D points in the tree buckets.

Efficiency of the Cached k-d Tree Search. Since k-d tree search and cached k-d tree search are very similar, most parts of the code were identical in the comparison experiments for k-d tree search vs. cached k-d tree search. In all tests, ICP with cached k-d tree search outperformed ICP with conventional k-d tree search. Three detailed analyses are provided:

1. The performance of the cached k-d tree search depending on a change of the bucket size was tested: For small bucket sizes, the speed-up is larger (Figure 4.17, top). This behaviour originates from the increasing time needed to search larger buckets.
2. The search time per iteration was recorded during the experiments (Figure 4.17, middle). For the first iteration the search times are equal, since cached k-d tree search uses conventional k-d tree search to create the cache. In the following iterations, the search time drops significantly and remains nearly constant. The conventional k-d tree search increases in speed, too. Here, the amount of backtracking is reduced due to the fact that the calculated transformations $(\boldsymbol{R}, \boldsymbol{t})$ are getting smaller.
3. The number of points to register influences the search time. With increasing number of points, the positive effect of caching algorithms becomes more and more significant (Figure 4.17, bottom).

Chapter 5
Globally Consistent Range Image Registration

5.1 Explicit Loop Closing

5.1.1 Extrapolating the Odometry to Six Degrees of Freedom

Since nearly all mobile robots have an odometer to measure traveled distances, our algorithm uses these measurements to calculate a first pose estimate. The odometry is extrapolated to 6 degrees of freedom using previous registration matrices. We are using a left-handed coordinate system, i.e., the y coordinate represents elevation. Then, the change of the robot pose $\boldsymbol{\Delta P}$ given the odometry information $(x_n^{\text{odo}}, z_n^{\text{odo}}, \theta_{y,n}^{\text{odo}})$, $(x_{n+1}^{\text{odo}}, z_{n+1}^{\text{odo}}, \theta_{y,n+1}^{\text{odo}})$ and the registration matrix $\boldsymbol{R}(\theta_{x,n}, \theta_{y,n}, \theta_{z,n})$, is calculated by solving:

$$\begin{pmatrix} x_{n+1}^{\text{odo}} \\ 0 \\ z_{n+1}^{\text{odo}} \\ 0 \\ \theta_{y,n+1}^{\text{odo}} \\ 0 \end{pmatrix} = \begin{pmatrix} x_{n}^{\text{odo}} \\ 0 \\ z_{n}^{\text{odo}} \\ 0 \\ \theta_{y,n}^{\text{odo}} \\ 0 \end{pmatrix} + \left(\begin{array}{c|ccc} \boldsymbol{R}(\theta_{x,n}, \theta_{y,n}, \theta_{z,n}) & & \mathbf{0} & \\ \hline & 1 & 0 & 0 \\ \mathbf{0} & 0 & 1 & 0 \\ & 0 & 0 & 1 \end{array}\right) \cdot \underbrace{\begin{pmatrix} \Delta x_{n+1} \\ \Delta y_{n+1} \\ \Delta z_{n+1} \\ \Delta\theta_{x,n+1} \\ \Delta\theta_{y,n+1} \\ \Delta\theta_{z,n+1} \end{pmatrix}}_{\boldsymbol{\Delta P}}.$$

Therefore, calculating $\boldsymbol{\Delta P} = (\Delta x_{n+1}, \Delta y_{n+1}, \Delta z_{n+1}, \Delta\theta_{x,n+1}, \Delta\theta_{y,n+1}, \Delta\theta_{z,n+1})$ requires a matrix inversion. If $n = 1$, $\boldsymbol{R}(\theta_{x,n}, \theta_{y,n})$ is set to the identity matrix. Finally, the 6D pose $\boldsymbol{P}_{n+1}$ is calculated by

$$\boldsymbol{P}_{n+1} = \boldsymbol{\Delta P} \cdot \boldsymbol{P}_n$$

A. Nüchter: 3D Robotic Map.: The Simultaneous Local. & Map. Prob., STAR 52, pp. 77–108.
springerlink.com

using the poses' matrix representations. Thus, the planar odometry is extrapolated in 6D to the (x, z)-plane defined by the last robot pose. Note that the values for y_{n+1}, $\theta_{x,n+1}$, and $\theta_{z,n+1}$ are usually not equal to zero, due to the matrix inversion.

5.1.2 Closed Loop Detection

An imporant issue of all SLAM algorithms is the detection of closed loops. Closed loops enable SLAM algorithms to deform the trajectory to make maps globally consistent. Their detection answers the question, if the robot were already at this position or at a position close by.

Visual Closed Loop Detection

Many approaches have been proposed to solve this key chellenge. Most robots use camera data to detect closed loops. Camera data for loop detection is utilized by extracting features from the captured images and processing them. This is normally done in three steps: Detecting the features, encoding them using feature descriptors, and finally matching them against each other. Many region detectors are available, namely, the MSER detector, the Salient region detector, the Harris-Affine detector, the Hessian-Affine detector, the Intensity extrema based detector (IBR), and the Edge based detector (EBR) [66, 80, 81, 109, 129]. There are also quite a number of feature descriptors, e.g., SIFT, GLOH, Shape Context, PCA, Moments, Cross correlation and Steerable filters [83]. Important attributes are invariance to scale, rotation, transformation or changes in illumination. A good overview of both detectors and descriptors is given in [82].

The visual loop closing procedure is basically designed as follows: Images are taken from the robot in an incremental fashion. These images are applied to the visual feature extraction pipeline one at a time. The results of the pipeline, the extracted features, are stored in a database. After the first image is processed, the resulting features of each succeeding image are matched against all features that are already stored in the database to detect a loop. The matching of the features is equivalent to the distance between vectors in a high dimensional vector space. A loop closing hypothesis is generated if similar features are found in two images, that is, if their distance is below a certain threshold [71].

Closed Loop Detection with Laser Scans

Detecting if a robot has already been at some pose based on 3D scans is a complex topic and includes among other things place recognition. However, having a decent pose estimate via odometry extrapolation and the ICP algorithm, one can confine closed loop detection to a comparison of scan poses.

We define a closed loop as detected, iff the robot position is withing a certain limit, e.g., 5 m. Then the scans should be matchable, i.e., a sufficient number of point pairs can be calculated.

If the 3D laser range finder employs an apex angle, i.e., covers only a part of the robot surrounding, the distance threshold criterion will fail. Therefore we include additionally the number of matched points into the closed loop detection, i.e., this number must exceed a certain threshold.

The methods above are only heuristics. In the remainder of this book, it will be shown, that for our 6D SLAM these heuristics are sufficient. Nevertheless, closed loop detection and place recognition remain a key challenge in mobile robotics.

5.1.3 Error Diffusion

After a loop has been detected the accumulated error has to be distributed over the scans forming the loop. The error consists of a rotational part $\boldsymbol{R}$ and a translation error $\boldsymbol{t}$ and is determined by matching the last scan of the loop against the first one. Fractional parts of this transformation have to be computed. Assume a loop consists of n 3D scans and the n-th was matched against scan number 1 yielding a transformation $(\boldsymbol{R}, \boldsymbol{t})$. The remaining scans of the loop are updated as follows:

1. The translational part is calculated in a straightfoward fashion.

$$\boldsymbol{t}_i = \frac{i}{n-1}\boldsymbol{t}.$$

2. Interpolation between rotation matrices is necessery to determine a fractional part of the rotation. Interpolation between rotation matrices is difficult, so we use the quaternion and axis angle representation here (cf. Appendix A.1). A rotation about an axis $\boldsymbol{a}$ and an angle θ is split up into fractions but must be calculated via a detour using quaternions [22, 56, 96]. Similarily to the equation (4.7) a rotation matrix $\boldsymbol{R}$ is transformed into a quaternion using the inverse calculation rule as follows (see also [28, 112]):

$$\dot{\boldsymbol{q}} = \begin{pmatrix} q_0 \\ q_x \\ q_y \\ q_z \end{pmatrix} = \begin{pmatrix} \frac{1}{2}\sqrt{1 + \operatorname{tr}(\boldsymbol{R})} \\ \frac{1}{2}\frac{1 + r_{3,3} - r_{3,2}}{\sqrt{\operatorname{tr}(\boldsymbol{R})}} \\ \frac{1}{2}\frac{1 + r_{2,1} - r_{2,3}}{\sqrt{\operatorname{tr}(\boldsymbol{R})}} \\ \frac{1}{2}\frac{1 + r_{1,2} - r_{1,1}}{\sqrt{\operatorname{tr}(\boldsymbol{R})}} \end{pmatrix}$$

where $r_{i,j}$ are the elements of $\boldsymbol{R}$. If the trace of the matrix $\operatorname{tr}(\boldsymbol{R})$ is zero the above calculation cannot be perfomed. In this case the following calculation rule applies: If $r_{1,1} > r_{2,2}$ and $r_{1,1} > r_{3,3}$ holds, then

$$\dot{\boldsymbol{q}} = \begin{pmatrix} \frac{1}{2}\frac{r_{2,3}-r_{3,2}}{\sqrt{1+r_{1,1}-r_{2,2}-r_{3,3}}} \\ \frac{1}{2}\sqrt{1+r_{1,1}-r_{2,2}-r_{3,3}} \\ \frac{1}{2}\frac{r_{1,2}+r_{2,1}}{\sqrt{1+r_{1,1}-r_{2,2}-r_{3,3}}} \\ \frac{1}{2}\frac{r_{3,1}+r_{1,3}}{\sqrt{1+r_{1,1}-r_{2,2}-r_{3,3}}} \end{pmatrix},$$

if $r_{2,2} > r_{3,3}$ holds, we have

$$\dot{\boldsymbol{q}} = \begin{pmatrix} \frac{1}{2}\frac{r_{3,1}-r_{1,3}}{\sqrt{1-r_{1,1}+r_{2,2}-r_{3,3}}} \\ \frac{1}{2}\frac{r_{1,2}+r_{2,1}}{\sqrt{1-r_{1,1}+r_{2,2}-r_{3,3}}} \\ \frac{1}{2}\sqrt{1-r_{1,1}+r_{2,2}-r_{3,3}} \\ \frac{1}{2}\frac{r_{2,3}+r_{3,2}}{\sqrt{1-r_{1,1}+r_{2,2}-r_{3,3}}} \end{pmatrix},$$

otherwise the quaternion $\dot{\boldsymbol{q}}$ is computed as follows:

$$\dot{\boldsymbol{q}} = \begin{pmatrix} \frac{1}{2}\frac{r_{1,2}-r_{2,1}}{\sqrt{1-r_{1,1}-r_{2,2}+r_{3,3}}} \\ \frac{1}{2}\frac{r_{3,1}+r_{1,3}}{\sqrt{1+r_{1,1}-r_{2,2}-r_{3,3}}} \\ \frac{1}{2}\frac{r_{2,3}+r_{3,2}}{\sqrt{1-r_{1,1}-r_{2,2}+r_{3,3}}} \\ \frac{1}{2}\sqrt{1-r_{1,1}-r_{2,2}+r_{3,3}} \end{pmatrix}.$$

After the rotation is converted into the corresponding quaternion, we construct accoring to Eq. (4.14) a rotation axis $\boldsymbol{a}$ and calculate the rotation angle θ. This is performed after normalizing the quaternion $\dot{\boldsymbol{q}}$ by

$$\boldsymbol{a} = \begin{pmatrix} \frac{q_x}{\sqrt{1-q_0^2}} \\ \frac{q_y}{\sqrt{1-q_0^2}} \\ \frac{q_z}{\sqrt{1-q_0^2}} \end{pmatrix} \quad \text{and} \quad \theta = 2\arccos q_o.$$

Now, the angle θ is distributed over all $(n-1)$ 3D scans of the loop and the interpolating matrices are given as:

$$\boldsymbol{R}_i = \begin{pmatrix} \cos\frac{i\theta}{n-1} + a_x^2(1-\cos\frac{i\theta}{n-1}) & a_z \sin\frac{i\theta}{n-1} + a_x a_y(1-\cos\frac{i\theta}{n-1}) & -a_y \sin\frac{i\theta}{n-1} + a_x a_z(1-\cos\frac{i\theta}{n-1}) \\ -a_z \sin\frac{i}{n-1}\theta + a_x a_y(1-\cos\frac{i}{n-1}\theta) & \cos\frac{i\theta}{n-1} + a_y^2(1-\cos\frac{i\theta}{n-1}) & -a_x \sin\frac{i\theta}{n-1} + a_y a_z(1-\cos\frac{i\theta}{n-1}) \\ a_y \sin\frac{i\theta}{n-1} + a_x a_z(1-\cos\frac{i\theta}{n-1}) & -a_x \sin\frac{i\theta}{n-1} + a_y a_z(1-\cos\frac{i\theta}{n-1}) & \cos\frac{i\theta}{n-1} + a_z^2(1-\cos\frac{i\theta}{n-1}) \end{pmatrix}.$$

Using the matrices $\boldsymbol{R}_i$ and the vectors $\boldsymbol{t}_i$ the robot poses of the loop are updated.

5.2 GraphSLAM as Generalization of Loop Closing

5.2.1 Algorithm Overview

In the following we describe a GraphSLAM method that considers in contrast to the previous explicit loop losing a graph connecting matchable poses. It is an extention to the scan matching procedure suggested by Lu and Milion 1997 [78], referred here as LUM. It employs a rotation representation using Euler angles. The scan matching process is outlined in Figure 5.1.

During the acquisition (or processing) of 3D scans odometry extrapolation and ICP scan matching yield a stable pose estimate. ICP iterates two steps: First, scan matching, i.e., computing closest points and second, minimization of the error function to yield a new pose estimate. After a scan has been processed with ICP, closed loops are detected and a graph of the pose relation is built. Finally global relaxation is started in an iterative fashion by perfoming scan matching again, i.e. computing closest points, linearization and solving a system of linear equations. This process continues until the whole trajectory is processed.

5.2.2 Probabilistic Global Relaxation Based on Scan Matching

To generate a globally consistent map from a set of relations, Lu and Milios formulated a probabilistic algorithm that considers all pose-estimations and their covariances at once [78].

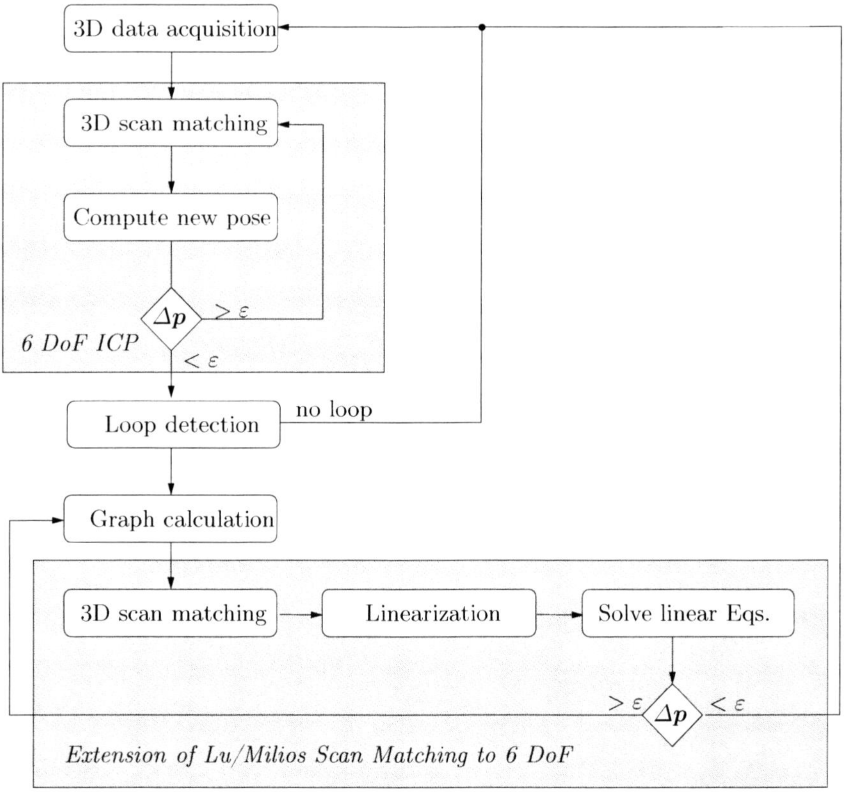

Fig. 5.1 The interaction of ICP and LUM in the overall algorithm. The threshold ε used to determine if a scan changed its pose is the same in both algorithm parts.

Problem Formulation

Consider a robot traveling along a path, and traversing the $n+1$ poses $\boldsymbol{P}_0, \ldots, \boldsymbol{P}_n$. At each pose $\boldsymbol{P}_i$, the robot takes a laser scan of its environment. By matching two scans made at two different poses, we acquire a set of neighbor relations between these poses. In the resulting network, nodes represent poses, and edges neighbor relations between them. Given such a network with $n+1$ graph nodes $\boldsymbol{X}_0, \ldots, \boldsymbol{X}_n$ and the directed edges $\boldsymbol{D}_{i,j}$, we want to estimate optimally all poses to build a consistent map of the environment. For simplification, the measurement equation is assumed to be linear:

$$\boldsymbol{D}_{i,j} = \boldsymbol{X}_i - \boldsymbol{X}_j.$$

The observation $\bar{\boldsymbol{D}}_{i,j}$ of the true underlying difference is modeled as $\bar{\boldsymbol{D}}_{i,j} = \boldsymbol{D}_{i,j} + \Delta\boldsymbol{D}_{i,j}$ where $\Delta\boldsymbol{D}_{i,j}$ is a Gaussian distributed error with zero mean and a covariance matrix $\boldsymbol{C}_{i,j}$, that is assumed to be known.

Maximum likelihood estimation is used to approximate the optimal poses $\boldsymbol{X}_i$. Under the assumption that all errors in the observations are Gaussian and independently distributed, maximizing the probability of all $\boldsymbol{D}_{i,j}$, given their observations $\bar{\boldsymbol{D}}_{i,j}$, is equivalent to minimizing the following Mahalanobis distance:

$$\mathbf{W} = \sum_{(i,j)} (\boldsymbol{D}_{i,j} - \bar{\boldsymbol{D}}_{i,j})^T \boldsymbol{C}_{i,j}^{-1} (\boldsymbol{D}_{i,j} - \bar{\boldsymbol{D}}_{i,j}). \tag{5.1}$$

Solution as given by Lu and Milios

We consider the simple linear case of the estimation problem. Without loss of generality we assume that the network is fully connected, i,e, each pair of nodes $\boldsymbol{X}_i$, $\boldsymbol{X}_j$ is connected by a link $\boldsymbol{D}_{i,j}$. In the case of a missing link $\boldsymbol{D}_{i,j}$ we set the corresponding $\boldsymbol{C}_{i,j}^{-1}$ to $\mathbf{0}$. Eq. (5.1) unfolds:

$$\mathbf{W} = \sum_{(0 \le i < j \le n)} (\boldsymbol{X}_i - \boldsymbol{X}_j - \bar{\boldsymbol{D}}_{i,j})^T \boldsymbol{C}_{i,j}^{-1} (\boldsymbol{X}_i - \boldsymbol{X}_j - \bar{\boldsymbol{D}}_{i,j}). \tag{5.2}$$

To minimize equation (5.2), a coordinate system is defined by setting one node as a reference point. Setting $\boldsymbol{X}_0 = (0,0,0,0,0,0)$, the n free nodes $\boldsymbol{X}_1, \ldots, \boldsymbol{X}_n$ denote the poses relative to $\boldsymbol{X}_0$.

Using the signed incidence matrix $\mathbf{H}$, the concatenated measurement equation $\mathbf{D}$ is written as

$$\mathbf{D} = \mathbf{H}\mathbf{X},$$

with the concatenation $\boldsymbol{X}$ of $\boldsymbol{X}_1$ to $\boldsymbol{X}_n$. The Mahalanobis distance equation (5.1) can be written as:

$$\mathbf{W} = (\bar{\mathbf{D}} - \mathbf{H}\mathbf{X})^T \mathbf{C}^{-1} (\bar{\mathbf{D}} - \mathbf{H}\mathbf{X}).$$

The concatenation of all observations $\bar{\boldsymbol{D}}_{i,j}$ forms the vector $\bar{\mathbf{D}}$, while $\mathbf{C}$ is a block-diagonal matrix comprised of $\boldsymbol{C}_{i,j}$ as submatrices. The solution $\mathbf{X}$ that minimizes the equation (5.2), and its covariance $\mathbf{C_X}$ are given by

$$\begin{aligned} \mathbf{X} &= (\mathbf{H}^T \mathbf{C}^{-1} \mathbf{H})^{-1} \mathbf{H}^T \mathbf{C}^{-1} \bar{\mathbf{D}} \\ \mathbf{C_X} &= (\mathbf{H}^T \mathbf{C}^{-1} \mathbf{H})^{-1}. \end{aligned} \tag{5.3}$$

The matrix $\mathbf{G} = \mathbf{H}^T \mathbf{C}^{-1} \mathbf{H}$ and the vector $\mathbf{B} = \mathbf{H}^T \mathbf{C}^{-1} \bar{\mathbf{D}}$ simplify the notation of the solution. $\mathbf{G}$ consists of submatrices

$$\boldsymbol{G}_{i,j} = \begin{cases} \sum_{j=0}^{n} \boldsymbol{C}_{i,j}^{-1} & (i = j) \\ \boldsymbol{C}_{i,j}^{-1} & (i \neq j). \end{cases} \tag{5.4}$$

The entries of $\mathbf{B}$ are obtained by:

$$\boldsymbol{B}_i = \sum_{\substack{j=0 \\ j \neq i}}^{n} \boldsymbol{C}_{i,j}^{-1} \bar{\boldsymbol{D}}_{i,j}. \tag{5.5}$$

Solving the linear optimal estimation problem (5.3) is equivalent to solving the following linear equation system:

$$\mathbf{G}\mathbf{X} = \mathbf{B}. \tag{5.6}$$

The proposed algorithm then requires $\mathbf{G}$ to be invertible. It can be shown [13] that this holds true when certain conditions are met. First, the covariance matrices must either be all positive definite or all negative definite. Secondly, the pose graph must be minimally connected, that is, there exists no two separated subgraphs, a condition trivially met in practice. In this case $\mathbf{G}$ is also positive definite respectively negative definite.

The Extension to Six Degrees of Freedom

To apply the above solution to mobile robots, it is necessary to linearize the pose difference equation (5.2). The three DoF case, i.e., $(x, z, \theta)^T$ poses, was solved by Lu and Milios [78]. In addition to pose relations from scan matching, Lu and Milios considered pose relations from odometry. Our algorithm derives relations for 6 DoF poses, i.e., $(x, y, z, \theta_x, \theta_y, \theta_z)^T$, by matching data obtained by a 3D laser range finder. The challenges of the extension were:

The amount of data. A 3D laser range finder scans the environment with a large number of samples. Typical resolutions in our robotics applications vary between 23168 and 360500 range points per 3D scan.

Linearization. Linearization of the rotation must regard the 3 DoF. The rotation consists of the three Euler angles $(\theta_x, \theta_y, \theta_z)$, and the multiplication of the corresponding three rotation matrices results in the desired overall rotation. By linearizing the Euler angles, we enforce valid rotation matrices.

Solution space. The additional three DoF result in an exponentially larger solution space. The solution is computationally more complex.

We define a 6D pose relation as follows: Assume that a robot starts at the pose $\boldsymbol{V}_b = (x_b, y_b, z_b, \theta_{x_b}, \theta_{y_b}, \theta_{z_b})^T$ and changes its pose by $\boldsymbol{D} = (x, y, z, \theta_x, \theta_y, \theta_z)^T$ relative to $\boldsymbol{V}_b$, ending up at $\boldsymbol{V}_a = (x_a, y_a, z_a, \theta_{x_a}, \theta_{y_a}, \theta_{z_a})^T$. The poses $\boldsymbol{V}_a$ and $\boldsymbol{V}_b$ are related by the compounding operation $\boldsymbol{V}_a = \boldsymbol{V}_b \oplus \boldsymbol{D}$. Similarly, a 3D position vector $\boldsymbol{u} = (x_u, y_u, z_u)$ is compounded with the pose $\boldsymbol{V}_b$ by $\boldsymbol{u}' = \boldsymbol{V}_b \oplus \boldsymbol{u}$:

$$
\begin{aligned}
x'_u &= x_b - z_u \sin\theta_{y_b} + \cos\theta_{y_b}(x_u \cos\theta_{z_b} - y_u \sin\theta_{z_b}) \\
&=: f_1(\boldsymbol{V}_b) \\
\\
y'_u &= y_b + z_u \cos\theta_{y_b} \sin\theta_{x_b} + \cos\theta_{x_b}(y_u \cos\theta_{z_b} + x_u \sin\theta_{z_b}) \\
&\quad + \sin\theta_{x_b} \sin\theta_{y_b}(x_u \cos\theta_{z_b} - y_u \sin\theta_{z_b}) \\
&=: f_2(\boldsymbol{V}_b) \\
\\
z'_u &= z_b - \sin\theta_{x_b}(y_u \cos\theta_{z_b} + x_u \sin\theta_{z_b}) \\
&\quad + \cos\theta_{x_b}(z_u \cos\theta_{y_b} + \sin\theta_{y_b}(x_u \cos\theta_{z_b} - y_u \sin\theta_{z_b})) \\
&=: f_3(\boldsymbol{V}_b)
\end{aligned}
\tag{5.7}
$$

This operation is used to transform a non-oriented point (from the scanner data) from its local to the global coordinate system.

Scan matching computes a set of m corresponding point pairs $\boldsymbol{u}_k^a, \boldsymbol{u}_k^b$ between two scans, each representing a single physical point. The positional error made by identifying these two points in different scans, is described by:

$$
\begin{aligned}
F_{ab}(\boldsymbol{V}_a, \boldsymbol{V}_b) &= \sum_{k=1}^{m} \left\| \boldsymbol{V}_a \oplus \boldsymbol{u}_k^a - \boldsymbol{V}_b \oplus \boldsymbol{u}_k^b \right\|^2 \\
&= \sum_{k=1}^{m} \left\| (\boldsymbol{V}_a \ominus \boldsymbol{V}_b) \oplus \boldsymbol{u}_k^a - \boldsymbol{u}_k^b \right\|^2 .
\end{aligned}
\tag{5.8}
$$

Based on these m point pairs, the algorithm computes the matrices $\bar{\boldsymbol{D}}_{i,j}$ and $\boldsymbol{C}_{i,j}$ for solving Eq. (5.1). $\bar{\boldsymbol{D}}_{i,j}$ is derived as follows:

Let $\bar{\boldsymbol{V}}_a = (\bar{x}_a, \bar{y}_a, \bar{z}_a, \bar{\theta}_{x_a}, \bar{\theta}_{y_a}, \bar{\theta}_{z_a})$ and $\bar{\boldsymbol{V}}_b = (\bar{x}_b, \bar{y}_b, \bar{z}_b, \bar{\theta}_{x_b}, \bar{\theta}_{y_b}, \bar{\theta}_{z_b})$ be close estimates of $\boldsymbol{V}_a$ and $\boldsymbol{V}_b$. If the global coordinates of a pair of matching points $\boldsymbol{u}_k = (x_k, y_k, z_k)$, then $(\boldsymbol{u}_k^a, \boldsymbol{u}_k^b)$ fulfills the equation

$$
\boldsymbol{u}_k \approx \boldsymbol{V}_a \oplus \boldsymbol{u}_k^a \approx \boldsymbol{V}_b \oplus \boldsymbol{u}_k^b .
$$

For small errors $\Delta\boldsymbol{V}_a = \bar{\boldsymbol{V}}_a - \boldsymbol{V}_a$ and $\Delta\boldsymbol{V}_b = \bar{\boldsymbol{V}}_b - \boldsymbol{V}_b$, a Taylor expansion leads to:

$$
\begin{aligned}
\Delta\boldsymbol{Z}_k &= \boldsymbol{V}_a \oplus \boldsymbol{u}_k^a - \boldsymbol{V}_b \oplus \boldsymbol{u}_k^b \\
&:= F_k(\boldsymbol{V}_a, \boldsymbol{V}_b) \\
&\approx F_k(\bar{\boldsymbol{V}}_a, \bar{\boldsymbol{V}}_b) - [\nabla_{\bar{\boldsymbol{V}}_a}(F_k(\bar{\boldsymbol{V}}_a, \bar{\boldsymbol{V}}_b))\Delta\boldsymbol{V}_a - \nabla_{\bar{\boldsymbol{V}}_b}(F_k(\bar{\boldsymbol{V}}_a, \bar{\boldsymbol{V}}_b))\Delta\boldsymbol{V}_b] \\
&= \bar{\boldsymbol{V}}_a \oplus \boldsymbol{u}_k^a - \bar{\boldsymbol{V}}_b \oplus \boldsymbol{u}_k^b - [\nabla_{\bar{\boldsymbol{V}}_a}(\bar{\boldsymbol{V}}_a \oplus \boldsymbol{u}_k^a)\Delta\boldsymbol{V}_a - \nabla_{\bar{\boldsymbol{V}}_b}(\bar{\boldsymbol{V}}_b \oplus \boldsymbol{u}_k^b)\Delta\boldsymbol{V}_b]
\end{aligned}
\tag{5.9}
$$

where $\nabla_{\bar{\boldsymbol{V}}_a}(F_k(\bar{\boldsymbol{V}}_a, \bar{\boldsymbol{V}}_b))$ is the gradient of the pose compounding operation. By matrix decomposition

$$\begin{aligned}\boldsymbol{M}_k\boldsymbol{H}_a &= \nabla_{\bar{\boldsymbol{V}}_a}(F_k(\bar{\boldsymbol{V}}_a,\bar{\boldsymbol{V}}_b))\\ \boldsymbol{M}_k\boldsymbol{H}_b &= \nabla_{\bar{\boldsymbol{V}}_b}(F_k(\bar{\boldsymbol{V}}_a,\bar{\boldsymbol{V}}_b)),\end{aligned}$$

Eq. (5.9) simplifies to:

$$\begin{aligned}\Delta\boldsymbol{Z}_k &\approx \bar{\boldsymbol{V}}_a\oplus\boldsymbol{u}_k^a-\bar{\boldsymbol{V}}_b\oplus\boldsymbol{u}_k^b-\boldsymbol{M}_k(\boldsymbol{H}_a\Delta\boldsymbol{V}_a-\boldsymbol{H}_b\Delta\boldsymbol{V}_b]\\ &= \bar{\boldsymbol{Z}}_k-\boldsymbol{M}_k\boldsymbol{D}\end{aligned}$$

with

$$\begin{aligned}\bar{\boldsymbol{Z}}_k &= \bar{\boldsymbol{V}}_a\oplus\boldsymbol{u}_k^a-\bar{\boldsymbol{V}}_b\oplus\boldsymbol{u}_k^b\\ \boldsymbol{D} &= (\boldsymbol{H}_a\Delta\boldsymbol{V}_a-\boldsymbol{H}_b\Delta\boldsymbol{V}_b)\\ \boldsymbol{M}_k &= \begin{pmatrix}1&0&0&0&-y_k&-z_k\\0&1&0&z_k&x_k&0\\0&0&1&-y_k&0&x_k\end{pmatrix}\end{aligned}\tag{5.10}$$

$\boldsymbol{H}_a$ as

$$\boldsymbol{H}_a=\begin{pmatrix}1&0&0&0&\bar{z}_a\cos(\bar{\theta}_{x_a})+\bar{y}_a\sin(\bar{\theta}_{x_a})&\bar{y}_a\cos(\bar{\theta}_{x_a})\cos(\bar{\theta}_{y_a})-\bar{z}_a\cos(\bar{\theta}_{y_a})\sin(\bar{\theta}_{x_a})\\0&1&0&-\bar{z}_a&-\bar{x}_a\sin(\bar{\theta}_{x_a})&-\bar{x}_a\cos(\bar{\theta}_{x_a})\cos(\bar{\theta}_{y_a})-\bar{z}_a\sin(\bar{\theta}_{y_a})\\0&0&1&\bar{y}_a&-\bar{x}_a\cos(\bar{\theta}_{x_a})&\bar{x}_a\cos(\bar{\theta}_{y_a})\sin(\bar{\theta}_{x_a})+\bar{y}_a\sin(\bar{\theta}_{y_a})\\0&0&0&1&0&\sin(\bar{\theta}_{y_a})\\0&0&0&0&\sin(\bar{\theta}_{x_a})&\cos(\bar{\theta}_{x_a})\cos(\bar{\theta}_{y_a})\\0&0&0&0&\cos(\bar{\theta}_{x_a})&-\cos(\bar{\theta}_{y_a})\sin(\bar{\theta}_{x_a})\end{pmatrix}.$$

$\boldsymbol{H}_b$ is given analogously. This matrix decomposition and the derivation of $\boldsymbol{H}_a$, $\boldsymbol{H}_b$ are the crucial steps in extending Lu and Milios style SLAM to 6 DoF.

$\boldsymbol{D}$ as defined by Eq. (5.10) is the new linearized measurement equation. To calculate both $\bar{\boldsymbol{D}}$ and $\boldsymbol{C}_{\boldsymbol{D}}$, the equation (5.8) is rewritten in matrix form

$$F_{ab}(\boldsymbol{D})\approx(\mathbf{Z}-\mathbf{M}\boldsymbol{D})^T(\mathbf{Z}-\mathbf{M}\boldsymbol{D}).$$

$\mathbf{M}$ is the concatenated matrix consisting of all $\boldsymbol{M}_k$'s, and $\mathbf{Z}$ the concatenated vector consisting of all $\boldsymbol{Z}_k$'s. The vector $\bar{\boldsymbol{D}}$ that minimizes F_{ab} is given by

$$\bar{\boldsymbol{D}}=(\mathbf{M}^T\mathbf{M})^{-1}\mathbf{M}^T\mathbf{Z}.\tag{5.11}$$

Since minimizing F_{ab} constitutes a linear least squares regression, we model the Gaussian distribution of the solution with mean $\bar{\boldsymbol{D}}$ and standard covariance estimation

$$\boldsymbol{C_D} = s^2(\mathbf{M}^T\mathbf{M}). \tag{5.12}$$

$\boldsymbol{s}^2$ is the unbiased estimate of the covariance of the identically, independently distributed errors of $\boldsymbol{Z}_k$, given by:

$$s^2 = (\boldsymbol{Z} - \mathbf{M}\bar{\boldsymbol{D}})^T(\mathbf{Z} - \mathbf{M}\bar{\boldsymbol{D}})/(2m-3) = \frac{F_{ab}(\bar{\boldsymbol{D}})}{2m-3}.$$

The error term W_{ab} corresponding to our pose relation is defined by:

$$W_{ab} = (\bar{\boldsymbol{D}} - \boldsymbol{D})^T \boldsymbol{C}_{\boldsymbol{D}}^{-1}(\bar{\boldsymbol{D}} - \boldsymbol{D}).$$

5.2.3 *Transforming the Solution*

Solving the linear equation (5.6) leads to an optimal estimate of the new measurement equation of $\boldsymbol{D}$ (Eq. (5.10)). To yield an optimal estimation of the robot poses, it is necessary to transform $\boldsymbol{D}$. By this optimal estimation, a set of solutions $\boldsymbol{X}_i = \boldsymbol{H}_i \Delta \boldsymbol{V}_i$ is computed, each corresponding to a node in the network. Assuming that the reference pose $\boldsymbol{V}_0 = 0$, the pose $\boldsymbol{V}_i$ and its covariance $\boldsymbol{C}_i$ are updated by:

$$\boldsymbol{V}_i = \bar{\boldsymbol{V}}_i - \boldsymbol{H}_i^{-1}\boldsymbol{X}_i,$$
$$\boldsymbol{C}_i = (\boldsymbol{H}_i^{-1})\boldsymbol{C}_i^{\boldsymbol{X}}(\boldsymbol{H}_i^{-1})^T.$$

If $\boldsymbol{V}_0$ is nonzero, the solutions have to be transformed by:

$$\boldsymbol{V}'_i = \boldsymbol{V}_0 \oplus \boldsymbol{V}_i$$
$$\boldsymbol{C}'_i = \boldsymbol{K}_0 \boldsymbol{C}_i \boldsymbol{K}_0^T$$

where

$$\boldsymbol{K}_0 = \begin{pmatrix} \boldsymbol{R}_{\theta_{x0},\theta_{y0},\theta_{z0}} & 0 \\ 0 & \mathbb{1}_{3\times 3} \end{pmatrix}$$

with a rotation matrix $\boldsymbol{R}_{\theta_{x0},\theta_{y0},\theta_{z0}}$.

The Algorithm

The optimal estimation algorithm is given as Algorithm 5.1. Iterative execution of Algorithm 5.1 yields a successive improvement of the global pose estimation. Step 3 is sped up by component-wise computation of $\mathbf{G}$ and $\mathbf{B}$.

Algorithm 5.1. Optimal estimation algorithm (LUM).

1. Compute the point correspondences $\boldsymbol{u}_k^a, \boldsymbol{u}_k^b$.
2. For any link (i,j) in the given graph compute the measurement vector $\bar{\boldsymbol{D}}_{ij}$ by Eq. (5.11) and its covariance $\boldsymbol{C}_{ij}$ by Eq. (5.12).
3. From all $\bar{\boldsymbol{D}}_{ij}$ and $\boldsymbol{C}_{ij}$ form the linear system $\mathbf{GX} = \mathbf{B}$, with $\boldsymbol{G}$ and $\boldsymbol{B}$ as given in Eq. (5.4) and (5.5) respectively.
4. Solve for $\mathbf{X}$
5. Update the poses and their covariances, as explained in Section 5.2.3.

The components $\boldsymbol{C}_{i,j}^{-1} = (\mathbf{M}^T\mathbf{M})/\boldsymbol{s}^2$ and $\boldsymbol{C}_{i,j}^{-1}\bar{\boldsymbol{D}}_{i,j} = (\mathbf{M}^T\mathbf{Z})/\boldsymbol{s}^2$ are expanded into simple summations, as shown in the appendix 5.2.3. The most expensive operation is solving the linear equation system $\mathbf{GX} = \mathbf{B}$. Since $\mathbf{G}$ is a positive definite, symmetric $(6n \times 6n)$ matrix, this is done by Cholesky decomposition in $\mathcal{O}(n^3)$.

Invertibility of G. The proposed algorithm depends on the invertibility of the matrix $\mathbf{G}$. The following theorem proves it.

Theorem 9. *The matrix* $\mathbf{G}$ *is invertable if the follwoing conditions are met:*
1. All covariances are positive or negative definite, and:
2. The pose graph is connected, i.e., there exist no two separate subgraphs.

Proof. The second condition is trivially met in practice, since all consecutive poses are linked. We prove that $\mathbf{G}$ is positive definite, and therefore invertible, using induction. In order to simplify the proof, we show that changing the reference pose does not change the positive definiteness of $\mathbf{G}$. Without loss of generality, $\mathbf{G}$ is a positive definite matrix of the form (5.4), with the reference node $\boldsymbol{X}_0$. Switching the reference node to $\boldsymbol{X}_i$ results in the matrix $\mathbf{G}'$. These two matrices are related by

$$\mathbf{G}' = \mathbb{1}_i \mathbf{G}$$

where $\mathbb{1}_i$ is an identity matrix of size $(dn \times dn)$ with a row of negative $(d \times d)$ identity matrices of the form:

$$\mathbb{1}_i = \begin{pmatrix} \mathbb{1}_{d(i-1)} & 0 & 0 \\ -\mathbb{1}_d & \dots & -\mathbb{1}_d \\ 0 & 0 & \mathbb{1}_{d(n-i+1)} \end{pmatrix}.$$

Multiplication with $\mathbb{1}_i$ corresponds to replacing the submatrices at (i,j) with the negative sum of all submatrices at row j. Since $\mathbb{1}_i$ is invertible, $\mathbf{G}'$ remains positive definite.

Induction base $k = n$: Assuming a graph with $n+1$ nodes and n links. The matrix $\mathbf{G}$ is transformed into the block diagonal matrix $\mathbf{G}'$, composed of covariance matrices by

$$\mathbf{G}' = \mathbb{1}_{\boldsymbol{D}}\mathbf{G}\mathbb{1}_{\boldsymbol{D}}^T,$$

with an upper-right triangular matrix $\mathbb{1}_{\boldsymbol{D}}$ of d-dimensional identity matrices

$$\mathbb{1}_{\boldsymbol{D}} = \begin{pmatrix} \mathbb{1}_d & \dots & \mathbb{1}_d \\ & \ddots & \vdots \\ 0 & & \mathbb{1}_d \end{pmatrix}.$$

Since $\mathbf{G}'$ is given by

$$\begin{aligned} \boldsymbol{G}'_{i,i} &= \boldsymbol{C}^{-1}_{i-1,i} \\ \boldsymbol{G}'_{i,j} &= 0 \qquad\qquad (i \neq j) \end{aligned}$$

and all covariances are positive definite, $\mathbf{G}'$ itself is positive definite. The same holds for $\mathbf{G}$, as $1_{\boldsymbol{D}}$ is invertible.

Inductive step $k \to k+1$: Let $\mathbf{G}$ be a positive definite matrix that corresponds to a graph with $n+1$ nodes and k links. An additional link between the nodes $\boldsymbol{X}_i$ and $\boldsymbol{X}_j$ is inserted, with positive definite covariance $\boldsymbol{C}_{i,j}$. Without restriction, $\boldsymbol{X}_i$ is the reference node of the given graph, since the reference pose is arbitrary. Thus, the resulting matrix $\mathbf{G}'$ is changed only at the $d \times d$ submatrix $\mathbf{G}'_{j,j}$:

$$\boldsymbol{G}'_{j,j} = \boldsymbol{G}_{j,j} + \boldsymbol{C}^{-1}_{i,j}.$$

In case $\boldsymbol{C}_{i,j}$ is positive definite, $\boldsymbol{G}'^{*}$ is positive definite, too, if and only if

$$\mathbf{X}^T\mathbf{G}'\mathbf{X} > 0 \qquad \mathbf{X} \in \mathbb{R}^{dn} \qquad \mathbf{X} \neq 0,$$

which is equivalent to

$$\sum_{k,l=1}^{n} \boldsymbol{X}_k^T \boldsymbol{G}'_{k,l} \boldsymbol{X}_l > 0, \tag{5.13}$$

where $\boldsymbol{X}_k$ are the d-dimensional subvectors of $\mathbf{X}$. Expanding Eq. (5.13) to

$$\begin{aligned} \sum_{k,l=1}^{n} \boldsymbol{X}_k^T \boldsymbol{G}'_{k,l} \boldsymbol{X}_l &= \boldsymbol{X}_j^T \boldsymbol{G}'_{j,j} \boldsymbol{X}_j + \sum_{\substack{k,l=1 \\ k\neq j\neq l}}^{n} \boldsymbol{X}_k^T \boldsymbol{G}_{k,l} \boldsymbol{X}_l \\ &= \boldsymbol{X}_j^T \boldsymbol{C}^{-1}_{i,j} \boldsymbol{X}_j + \sum_{k,l=1}^{n} \boldsymbol{X}_k^T \boldsymbol{G}_{k,l} \boldsymbol{X}_l \\ &= \boldsymbol{X}_j^T \boldsymbol{C}^{-1}_{i,j} \boldsymbol{X}_j + \mathbf{X}^T \mathbf{G} \mathbf{X} > 0. \end{aligned}$$

Thus, $\mathbf{G}'$ is a positive definite matrix. In case of $\boldsymbol{C}_{i,j}$ being negative definite, the proof follows analogously. □

Performace Issues

The large amount of 3D data to be processed makes computing time an issue in globally consistent range scan matching. Our algorithm benefits from the network structure. Each scan has to be aligned only to few neighbors in the graph. Compared to ICP metascan matching, LUM becomes more advantageous with increasing number of scans n. Our SLAM algorithm spends $\mathcal{O}(n^3)$ time on matrix computation. The matrices $\mathbf{B}$ and $\mathbf{G}$ are filled efficiently using simple additions. However, calculating the corresponding points for a link needs $\mathcal{O}(N \log N)$, the methods discussed in Section 4.1.4. N denotes the number of points per 3D scan, $n \ll N$. In all experiments the most computing time was spent in step 2 of algorithm 5.1, e.g., $n < 13$, $N < 300000$ or $n < 468$, $N < 18000$, respectively (cf. Section 6.2 and Section 6.3). Due to these performace issues, the most fundametal issues are speed-ups for closest point computation in the scan matching context.

Fast Construction of the Linear Equation System. To solve the linear equation system $\mathbf{GX} = \mathbf{B}$,

$$\begin{aligned} \boldsymbol{C}_{\boldsymbol{D}}^{-1} &= (\mathbf{M}^T\mathbf{M})/s^2 \\ \boldsymbol{C}_{\boldsymbol{D}}^{-1}\bar{\boldsymbol{D}} &= (\mathbf{M}^T\mathbf{Z})/s^2. \end{aligned}$$

are needed. To calculate these efficiently, summations are substituted for matrix multiplication by using the regularities in the matrix $\mathbf{M}$. $\mathbf{M}^T\mathbf{M}$ is represented as a sum over all corresponding point pairs:

$$\mathbf{M}^T\mathbf{M} = \sum_{k=1}^{m} \begin{pmatrix} 1 & 0 & 0 & 0 & -y_k & -z_k \\ 0 & 1 & 0 & z_k & x_k & 0 \\ 0 & 0 & 1 & -y_k & 0 & x_k \\ 0 & z_k & -y_k & y_k^2 + z_k^2 & x_k z_k & -x_k y_k \\ -y_k & x_k & 0 & x_k z_k & y_k^2 + x_k^2 & y_k z_k \\ -z_k & 0 & x_k & -x_k y_k & y_k z_k & x_k^2 + z_k^2 \end{pmatrix}.$$

Similarly, $\mathbf{M}^T\mathbf{Z}$ is calculated as follows:

$$\mathbf{M}^T\mathbf{Z} = \sum_{k=0}^{m} \begin{pmatrix} \Delta x_k \\ \Delta y_k \\ \Delta z_k \\ -z_k \cdot \Delta y_k + y_k \cdot \Delta z_k \\ -y_k \cdot \Delta x_k + x_k \cdot \Delta y_k \\ z_k \cdot \Delta x_k - x_k \cdot \Delta z_k \end{pmatrix}$$

with

$$\begin{pmatrix} \Delta x_k \\ \Delta y_k \\ \Delta z_k \end{pmatrix} = \bar{\boldsymbol{Z}}_k = \bar{\boldsymbol{V}}_a \oplus \boldsymbol{u}_k^a - \bar{\boldsymbol{V}}_b \oplus \boldsymbol{u}_k^b$$

and an approximation for each point:

$$\begin{pmatrix} *1cx_k \\ y_k \\ z_k \end{pmatrix} = \boldsymbol{u}_k \approx (\bar{\boldsymbol{V}}_a \oplus \boldsymbol{u}_k^a + \bar{\boldsymbol{V}}_b \oplus \boldsymbol{u}_k^b)/2.$$

Finally, s^2 is a simple summation using the observation of the linearized measurement equation $\bar{\boldsymbol{D}} = (\mathbf{M}^T\mathbf{M})^{-1}\mathbf{M}^T\mathbf{Z}$:

$$\begin{aligned} \boldsymbol{s}^2 = \sum_{k=0}^{m} \Big[& (\Delta x_k - (\bar{D}_0 - y_k \cdot \bar{D}_4 + z_k \cdot \bar{D}_5))^2 \\ & + (\Delta y_k - (\bar{D}_1 - z_k \cdot \bar{D}_3 + x_k \cdot \bar{D}_4))^2 \\ & + (\Delta z_k - (\bar{D}_2 + y_k \cdot \bar{D}_3 - x_k \cdot \bar{D}_5))^2 \Big]. \end{aligned}$$

$\bar{D}_i$ denotes the i-th entry of the vector $\bar{\boldsymbol{D}}$. Summation of $\boldsymbol{C}_{\boldsymbol{D}}^{-1}$ and $\boldsymbol{C}_{\boldsymbol{D}}^{-1}\bar{\boldsymbol{D}}$ yields $\mathbf{B}$ and $\mathbf{G}$.

Exploiting Sparse Computations. Again our algorithm benefits from the network structure. Each scan has to be aligned to only few neighbors in the graph. Links exist between consecutive scans in the robot's path and additionally those scans that are spatially close. Consequently, most entries of matrix $\mathbf{G}$ are zero, e.g., $\mathbf{G}$ is sparse (cf. Figure 5.2). Since $\mathbf{G}$ is also positive definite, we apply a sparse Cholesky decomposition to speed up the matrix inversion [24]. Alternative approaches are described in [37, 93].

Sparse Cholesky Decomposition: By symbolic analysis of the non-zero pattern of matrix $\mathbf{G}$, a fill-reducing permutation $\boldsymbol{P}$ and an elimination tree are calculated. We use the minimum degree algorithm [25], a heuristic brute force method, to determine the fill-reducing permutation, e.g., a matrix $\boldsymbol{P}$ for which the factorization $\boldsymbol{P}\mathbf{G}\boldsymbol{P}^T$ contains a minimal number of non-zero entries. The elimination tree, a pruned form of the graph $\mathbf{G}_L$ that is equivalent to the Cholesky decomposition $\boldsymbol{L}$, gives the non-zero pattern of $\boldsymbol{L}$ (cf. Figure 5.3). The sparse system $\mathbf{GX} = \mathbf{B}$ becomes

$$\boldsymbol{P}\mathbf{G}\boldsymbol{P}^T\boldsymbol{P}\mathbf{X} = \boldsymbol{P}\mathbf{B},$$

that is solved efficiently with the Cholesky factorisation $\boldsymbol{L}\boldsymbol{L}^T = \boldsymbol{P}\mathbf{G}\boldsymbol{P}^T$ [25]. After solving the equation system $\boldsymbol{L}\boldsymbol{y} = \boldsymbol{P}\mathbf{B}$ for $\boldsymbol{y}$, $\boldsymbol{L}^T\boldsymbol{z} = \boldsymbol{y}$ is solved for $\boldsymbol{z}$, resulting in the solution $\mathbf{X} = \boldsymbol{P}^T\boldsymbol{z}$. Sparse Cholesky factorisation is done in $\mathcal{O}$(FLOPS), i.e., is linear in the number of graph edges.

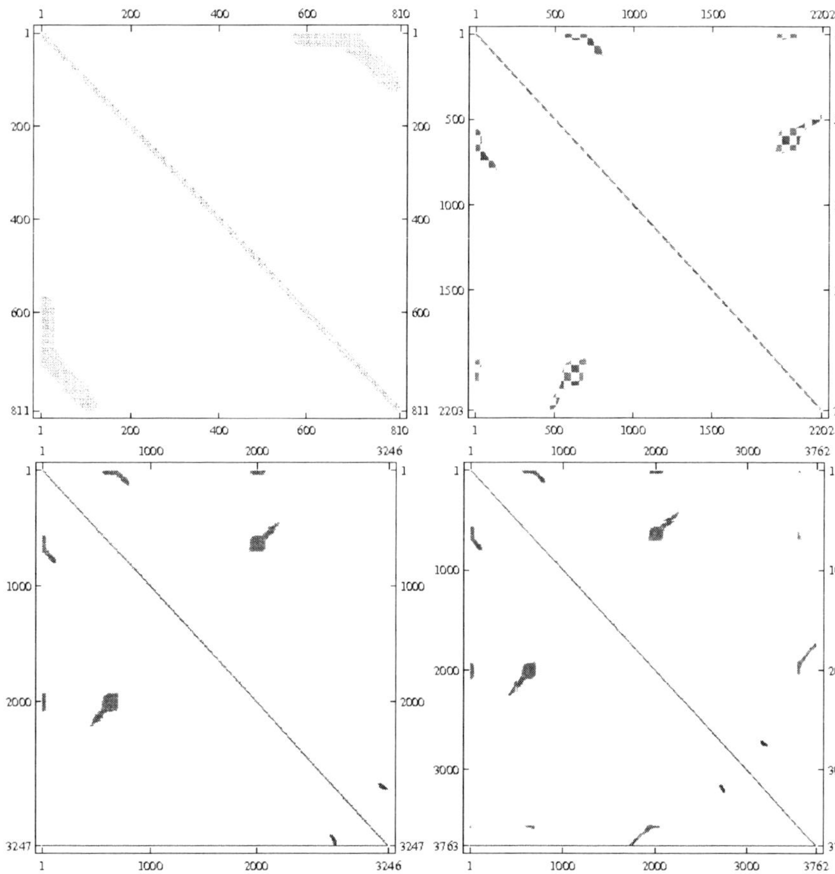

Fig. 5.2 Sparse matrices **G** generated for the Urban Environments data set (cf. Section 6.3) with an increasing number of scan poses. Different grey values correspond to negative and positive entries. White spaces are filled with zeros.

Fast Computation of Correspondences. According to our experience, the most computing time was spent in step 2 of Algorithm 5.1. While our matching algorithm spends $\mathcal{O}(n)$ time on matrix computation, calculating the corresponding points for a link needs $\mathcal{O}(N \log N)$, using cached k-d tree search [91]. N denotes the number of points per 3D scan, $n \ll N$. To avoid recomputing the k-d trees in each iteration, the query point is transformed into the local coordinate system according to the current scan pose. By this means, the k-d trees only have to be computed once at the beginning of global optimization, caching proceeds as described Sec. 4.1.4.

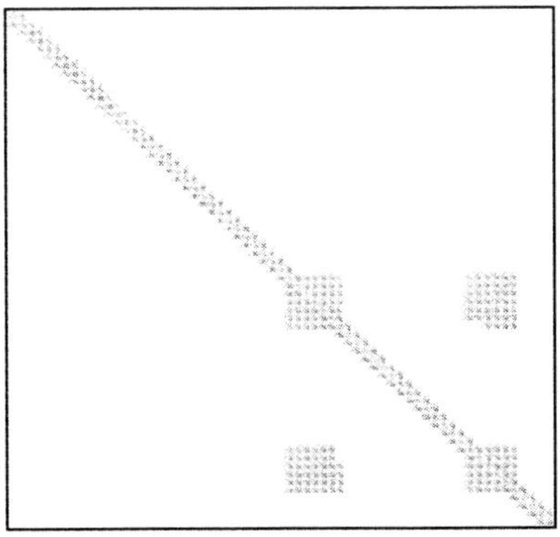

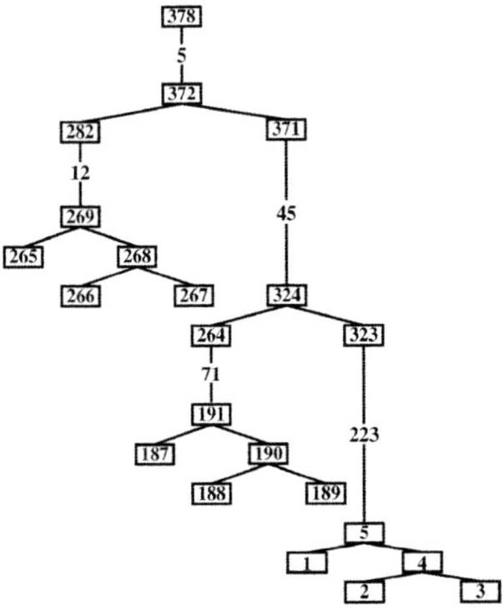

Fig. 5.3 Sparse Matrix **G** (left) and its elimination tree generated for the University building data set (cf. Sec. 6.1.3, Figure 6.6). The numbers on the edges of the tree indicate the number of linearly arranged nodes, removed to improve visibility. The nodes represent the columns of the matrix.

By evaluating the links in the graph, i.e., computing the entries of **G**, in a parallel fashion, the Algorithm 5.1 can be further accelerated.

5.3 Other Modeling Approaches

5.3.1 Linearization Using Quaternions

In the previous section the definition of the compound operator succeeded, since the rotation using Euler angles is linearized. To be precise, the Eq. (5.7) is linearized as follows and the non-trivial entries of $\nabla_{\bar{\boldsymbol{V}}_a}(\bar{\boldsymbol{V}}_a \oplus \boldsymbol{u}_k^a)$, i.e., the partial derivatives are:

$$
\begin{aligned}
\frac{\partial f_1}{\partial \theta_{x_a}}(\bar{\boldsymbol{V}}_a) &= 0 \\
\frac{\partial f_2}{\partial \theta_{x_a}}(\bar{\boldsymbol{V}}_a) &= -\sin\bar{\theta}_{x_a}(y_k^a \cos\bar{\theta}_{z_a} + x_k^a \sin\bar{\theta}_{z_a}) \\
&\quad + \cos\bar{\theta}_{x_a}(z_k^a \cos\bar{\theta}_{y_a} + \sin\bar{\theta}_{y_a}(x_k^a \cos\bar{\theta}_{z_a} - y_k^a \sin\bar{\theta}_{z_a})) \\
\frac{\partial f_3}{\partial \theta_{x_a}}(\bar{\boldsymbol{V}}_a) &= -\cos\bar{\theta}_{x_a}(y_k^a \cos\bar{\theta}_{z_a} + x_k^a \sin\bar{\theta}_{z_a}) \\
&\quad - \sin\bar{\theta}_{x_a}(z_k^a \cos\bar{\theta}_{y_a} + \sin\bar{\theta}_{y_a}(x_k^a \cos\bar{\theta}_{z_a} - y_k^a \sin\bar{\theta}_{z_a}))
\end{aligned}
$$

$$\frac{\partial f_1}{\partial \theta_{y_a}}(\bar{\boldsymbol{V}}_a) = -z_k^a \cos\bar{\theta}_{y_a} + \sin\bar{\theta}_{y_a}(y_k^a \sin\bar{\theta}_{z_a} - x_k^a \cos\bar{\theta}_{z_a})$$
$$\frac{\partial f_2}{\partial \theta_{y_a}}(\bar{\boldsymbol{V}}_a) = -\sin\bar{\theta}_{x_a}(z_k^a \sin\bar{\theta}_{y_a} + \cos\bar{\theta}_{y_a}(y_k^a \sin\bar{\theta}_{z_a}) - x_k^a \cos\bar{\theta}_{z_a})$$
$$\frac{\partial f_3}{\partial \theta_{y_a}}(\bar{\boldsymbol{V}}_a) = -\cos\bar{\theta}_{x_a}(z_k^a \sin\bar{\theta}_{y_a} + \cos\bar{\theta}_{y_a}(y_k^a \sin\bar{\theta}_{z_a}) - x_k^a \cos\bar{\theta}_{z_a})$$

$$\frac{\partial f_1}{\partial \theta_{z_a}}(\bar{\boldsymbol{V}}_a) = -\cos\bar{\theta}_{y_a}(y_k^a \cos\bar{\theta}_{z_a} + x_k^a \sin\bar{\theta}_{z_a})$$
$$\frac{\partial f_2}{\partial \theta_{z_a}}(\bar{\boldsymbol{V}}_a) = -\sin\bar{\theta}_{x_a}\sin\bar{\theta}_{y_a}(y_k^a \cos\bar{\theta}_{z_a} + x_k^a \sin\bar{\theta}_{z_a}) + \cos\bar{\theta}_{x_a}(x_k^a \cos\bar{\theta}_{z_a} - y_k^a \sin\bar{\theta}_{z_a})$$
$$\frac{\partial f_3}{\partial \theta_{z_a}}(\bar{\boldsymbol{V}}_a) = -\cos\bar{\theta}_{z_a}(x_k^a \sin\bar{\theta}_{x_a} + y_k^a \cos\bar{\theta}_{x_a}\sin\bar{\theta}_{y_a}) + (y_k^a \sin\bar{\theta}_{x_a} - x_k^a \cos\bar{\theta}_{x_a}\sin\bar{\theta}_{y_a})\sin\bar{\theta}_{z_a}.$$

Simple substitutions lead to the following form and all x_k^a,y_k^a and z_k^a vanish:

$$\frac{\partial f_1}{\partial \theta_{x_a}}(\bar{\boldsymbol{V}}_a) = 0$$
$$\frac{\partial f_2}{\partial \theta_{x_a}}(\bar{\boldsymbol{V}}_a) = z_k - \bar{z}_a$$
$$\frac{\partial f_3}{\partial \theta_{x_a}}(\bar{\boldsymbol{V}}_a) = \bar{y}_a - y_k$$

$$\frac{\partial f_1}{\partial \theta_{y_a}}(\bar{\boldsymbol{V}}_a) = (\bar{z}_a - z_k)\cos\bar{\theta}_{x_a} + (\bar{y}_a - y_k)\sin\bar{\theta}_{x_a}$$
$$\frac{\partial f_2}{\partial \theta_{y_a}}(\bar{\boldsymbol{V}}_a) = (x_k - \bar{x}_a)\sin\bar{\theta}_{x_a}$$
$$\frac{\partial f_3}{\partial \theta_{y_a}}(\bar{\boldsymbol{V}}_a) = (x_k - \bar{x}_a)\cos\bar{\theta}_{x_a}$$

$$\frac{\partial f_1}{\partial \theta_{z_a}}(\bar{\boldsymbol{V}}_a) = \cos\bar{\theta}_{y_a}((\bar{y}_a - y_k)\cos\bar{\theta}_{x_a} + (z_k - \bar{z}_a)\sin\bar{\theta}_{x_a}$$
$$\frac{\partial f_2}{\partial \theta_{z_a}}(\bar{\boldsymbol{V}}_a) = (x_k - \bar{x}_a)\cos\bar{\theta}_{x_a}\cos\bar{\theta}_{y_a} + (z_k - \bar{z}_a)\sin\bar{\theta}_{y_a}$$
$$\frac{\partial f_3}{\partial \theta_{z_a}}(\bar{\boldsymbol{V}}_a) = (\bar{x}_a - x_k)\cos\bar{\theta}_{y_a}\sin\bar{\theta}_{x_a} + (\bar{y}_a - y_k)\sin\bar{\theta}_{y_a}.$$

Therefore $\nabla_{\bar{\boldsymbol{V}}_a}(\bar{\boldsymbol{V}}_a \oplus \boldsymbol{u}_k^a)$ can be written as:

$$\nabla_{\bar{\boldsymbol{V}}_a}(\bar{\boldsymbol{V}}_a \oplus \boldsymbol{u}_k^a) = \begin{pmatrix} 1 & 0 & 0 & \frac{\partial f_1}{\partial \theta_{x_a}}(\bar{\boldsymbol{V}}_a) & \frac{\partial f_1}{\partial \theta_{y_a}}(\bar{\boldsymbol{V}}_a) & \frac{\partial f_1}{\partial \theta_{z_a}}(\bar{\boldsymbol{V}}_a) \\ 0 & 1 & 0 & \frac{\partial f_2}{\partial \theta_{x_a}}(\bar{\boldsymbol{V}}_a) & \frac{\partial f_2}{\partial \theta_{y_a}}(\bar{\boldsymbol{V}}_a) & \frac{\partial f_2}{\partial \theta_{z_a}}(\bar{\boldsymbol{V}}_a) \\ 0 & 0 & 1 & \frac{\partial f_3}{\partial \theta_{x_a}}(\bar{\boldsymbol{V}}_a) & \frac{\partial f_3}{\partial \theta_{y_a}}(\bar{\boldsymbol{V}}_a) & \frac{\partial f_3}{\partial \theta_{z_a}}(\bar{\boldsymbol{V}}_a) \end{pmatrix}.$$

Based on this the matrix decomposition has been found that separates x_k,y_k and z_k from the entries of the pose estimation $\bar{\boldsymbol{V}}_a$.

The quaternion based approach is similar. Suppose the pose is defined using quaternions, i.e., $\boldsymbol{V}_a = (x_a, y_a, z_a, \dot{\boldsymbol{q}}_a)^T = (x_a, y_a, z_a, q_{a,0}, q_{a,x}, q_{a,y}, q_{a,z})^T$ and $\boldsymbol{V}_b = (x_b, y_b, z_b, \dot{\boldsymbol{q}}_b)^T = (x_b, y_b, z_b, q_{b,0}, q_{b,x}, q_{b,y}, q_{b,z})^T$ and their estimates are $\bar{\boldsymbol{V}}_a = (\bar{x}_a, \bar{y}_a, \bar{z}_a, \bar{\dot{\boldsymbol{q}}}_a)^T = (\bar{x}_a, \bar{y}_a, \bar{z}_a, \bar{q}_{a,0}, \bar{q}_{a,x}, \bar{q}_{a,y}, \bar{q}_{a,z})^T$ and $\bar{\boldsymbol{V}}_b = (\bar{x}_b, \bar{y}_b, \bar{z}_b, \bar{\dot{\boldsymbol{q}}}_b)^T = (\bar{x}_b, \bar{y}_b, \bar{z}_b, \bar{q}_{b,0}, \bar{q}_{b,x}, \bar{q}_{b,y}, \bar{q}_{b,z})^T$. Additionally we have again m 3D point correspondences $(\boldsymbol{u}_k^a, \boldsymbol{u}_k^b)$ where $\boldsymbol{u}_k^a = (x_k^a, y_k^a, z_k^a)$ and $\boldsymbol{u}_k^b = (x_k^b, y_k^b, z_k^b)$, that correspond to the m points $\boldsymbol{u}_k = (x_k, y_k, z_k)$. Then we derive:

$$\begin{aligned}
\boldsymbol{u}_k &= \boldsymbol{V}_a \oplus \boldsymbol{u}_k^a \\
x_k &= x_a + (q_{a,0}^2 + q_{a,x}^2 - q_{a,y}^2 - q_{a,z}^2)x_k^a + 2(q_{a,x}^2 q_{a,y}^2 - q_{a,0}^2 q_{a,z}^2)y_k^a \\
&\quad + 2(q_{a,0}^2 q_{a,y}^2 + q_{a,x}^2 q_{a,z}^2)z_k^a \\
&=: f_1(V_a) \\
y_k &= y_a + 2(q_{a,0}^2 q_{a,y}^2 + q_{a,0}^2 q_{a,z}^2)x_k^a + (q_{a,0}^2 - q_{a,x}^2 + q_{a,y}^2 - q_{a,z}^2)y_k^a \\
&\quad + 2(q_{a,y}^2 q_{a,z}^2 - q_{a,0}^2 q_{a,y}^2)z_k^a \\
&=: f_2(V_a) \\
z_k &= z_a + 2(q_{a,0}^2 q_{a,z}^2 - q_{a,0}^2 q_{a,y}^2)x_k^a + 2(q_{a,0}^2 q_{a,x}^2 + q_{a,y}^2 q_{a,z}^2)y_k^a \\
&\quad + (q_{a,0}^2 - q_{a,x}^2 - q_{a,y}^2 + q_{a,z}^2)z_k^a \\
&=: f_3(V_a)
\end{aligned}$$

After removing x_k^a,y_k^a and z_k^a, the interesting partial derivatives are:

$$\begin{aligned}
\frac{\partial f_1}{\partial q_{a,0}}(\bar{\boldsymbol{V}}_a) &= 2\bar{q}_{a,0}x_k + 2\bar{q}_{a,z}y_k - 2\bar{q}_{a,y}z_k - 2(\bar{p}_a\bar{x}_a + \bar{q}_{a,z}\bar{y}_a - \bar{q}_{a,y}\bar{z}_a) \\
\frac{\partial f_2}{\partial q_{a,0}}(\bar{\boldsymbol{V}}_a) &= -2\bar{q}_{a,z}x_k + 2\bar{q}_{a,0}y_k - 2\bar{q}_{a,x}z_k - 2(\bar{q}_{a,z}\bar{x}_a + \bar{q}_{a,0}\bar{y}_a - \bar{q}_{a,x}\bar{z}_a) \\
\frac{\partial f_3}{\partial q_{a,0}}(\bar{\boldsymbol{V}}_a) &= 2\bar{q}_{a,y}x_k - 2\bar{q}_{a,x}y_k + 2\bar{q}_{a,0}z_k - 2(\bar{q}_{a,y}\bar{x}_a - \bar{q}_{a,x}\bar{y}_a + \bar{q}_{a,0}\bar{z}_a)
\end{aligned}$$

$$\begin{aligned}
\frac{\partial f_1}{\partial q_{a,x}}(\bar{\boldsymbol{V}}_a) &= 2\bar{q}_{a,x}x_k + 2\bar{q}_{a,y}y_k + 2\bar{q}_{a,z}z_k - 2(\bar{q}_{a,x}\bar{x}_a + \bar{q}_{a,y}\bar{y}_a + \bar{q}_{a,z}\bar{z}_a) \\
\frac{\partial f_2}{\partial q_{a,x}}(\bar{\boldsymbol{V}}_a) &= -2\bar{q}_{a,y}x_k + 2\bar{q}_{a,x}y_k - 2\bar{q}_{a,0}z_k - 2(-\bar{q}_{a,y}\bar{x}_a + \bar{q}_{a,x}\bar{y}_a - \bar{q}_{a,0}\bar{z}_a) \\
\frac{\partial f_3}{\partial q_{a,x}}(\bar{\boldsymbol{V}}_a) &= -2\bar{q}_{a_z}x_k + 2\bar{q}_{a,0}y_k + 2\bar{q}_{a,x}z_k - 2(-\bar{q}_{a,z}\bar{x}_a + \bar{q}_{a,0}\bar{y}_a + \bar{q}_{a,x}\bar{z}_a)
\end{aligned}$$

$$\frac{\partial f_1}{\partial q_{a,y}}(\bar{\boldsymbol{V}}_a) = 2\bar{q}_{a,y}x_k - 2\bar{q}_{a,x}y_k + 2\bar{q}_{a,0}z_k - 2(\bar{q}_{a,y}\bar{x}_a - \bar{q}_{a,x}\bar{y}_a + \bar{q}_{a,0}\bar{z}_a)$$
$$\frac{\partial f_2}{\partial q_{a,y}}(\bar{\boldsymbol{V}}_a) = 2\bar{q}_{a,x}x_k + 2\bar{q}_{a,y}y_k + 2\bar{q}_{a,z}z_k - 2(\bar{q}_{a,x}\bar{x}_a + \bar{q}_{a,y}\bar{y}_a + \bar{q}_{a,z}\bar{z}_a)$$
$$\frac{\partial f_3}{\partial q_{a,x}}(\bar{\boldsymbol{V}}_a) = -2\bar{q}_{a,0}x_k - 2\bar{q}_{a,z}y_k + 2\bar{q}_{a,y}z_k - 2(-\bar{q}_{a,0}\bar{x}_a - \bar{q}_{a,z}\bar{y}_a + \bar{q}_{a,y}\bar{z}_a)$$

$$\frac{\partial f_1}{\partial q_{a,z}}(\bar{\boldsymbol{V}}_a) = 2\bar{q}_{a,z}x_k - 2\bar{q}_{a,0}y_k - 2\bar{q}_{a,x}z_k - 2(\bar{q}_{a,z}\bar{x}_a - \bar{q}_{a,0}\bar{y}_a - \bar{q}_{a,x}\bar{z}_a)$$
$$\frac{\partial f_2}{\partial q_{a,z}}(\bar{\boldsymbol{V}}_a) = 2\bar{q}_{a,0}x_k + 2\bar{q}_{a,z}y_k - 2\bar{q}_{a,y}z_k - 2(\bar{q}_{a,0}\bar{x}_a + \bar{q}_{a,z}\bar{y}_a - \bar{q}_{a,y}\bar{z}_a)$$

$$\frac{\partial f_3}{\partial q_{a,z}}(\bar{\boldsymbol{V}}_a) = 2\bar{q}_{a,x}x_k + 2\bar{q}_{a,y}y_k + 2\bar{q}_{a,z}z_k - 2(\bar{q}_{a,x}\bar{x}_a + \bar{q}_{a,y}\bar{y}_a + \bar{q}_{a,y}\bar{z}_a).$$

Using the same principle as we have used in the Euler angle case the matrix decomposition $\boldsymbol{M}_k\boldsymbol{H}_a$ and $\boldsymbol{M}_k\boldsymbol{H}_b$ are given as:

$$\boldsymbol{M}_k = \begin{pmatrix} 1 & 0 & 0 & x_k & 0 & -z_k & y_k \\ 0 & 1 & 0 & y_k & z_k & 0 & -x_k \\ 0 & 0 & 1 & z_k & -y_k & x_k & 0 \end{pmatrix}$$

$$\boldsymbol{H}_a = \begin{pmatrix} 1 & 0 & 0 & (\bar{q}_{a,y}\bar{z}_a - \bar{q}_{a,0}\bar{x}_a - \bar{q}_{a,z}\bar{y}_a) & -2(\bar{q}_{a,x}\bar{x}_a + \bar{q}_{a,y}\bar{y}_a + \bar{q}_{a,z}\bar{z}_a) \\ 0 & 1 & 0 & 2(\bar{q}_{a,x}\bar{z}_a - \bar{q}_{a,z}\bar{x}_a - \bar{q}_{a,0}\bar{y}_a) & (\bar{q}_{a,y}\bar{x}_a - \bar{q}_{a,x}\bar{y}_a + \bar{q}_{a,0}\bar{z}_a) \\ 0 & 0 & 1 & 2(\bar{q}_{a,x}\bar{y}_a - \bar{q}_{a,y}\bar{x}_a - \bar{q}_{a,0}\bar{z}_a) & 2(\bar{q}_{a,z}\bar{x}_a - \bar{q}_{a,0}\bar{y}_a - \bar{q}_{a,x}\bar{z}_a) \\ 0 & 0 & 0 & 2\bar{q}_{a,0} & 2\bar{q}_{a,x} \\ 0 & 0 & 0 & 2\bar{q}_{a,x} & -2\bar{q}_{a,0} \\ 0 & 0 & 0 & 2\bar{q}_{a,y} & -2\bar{q}_{a,z} \\ 0 & 0 & 0 & 2\bar{q}_{a,z} & 2\bar{q}_{a,y} \end{pmatrix.}$$
$$\begin{matrix} 2(\bar{q}_{a,x}\bar{y}_a - \bar{q}_{a,y}\bar{x}_a - \bar{q}_{a,0}\bar{z}_a) & 2(\bar{q}_{a,0}\bar{y}_a - \bar{q}_{a,z}\bar{x}_a + \bar{q}_{a,x}\bar{z}_a) \\ -2(\bar{q}_{a,x}\bar{x}_a + \bar{q}_{a,y}\bar{y}_a + \bar{q}_{a,z}\bar{z}_a) & 2(\bar{q}_{a,y}\bar{z}_a - \bar{q}_{a,0}\bar{x}_a - \bar{q}_{a,z}\bar{y}_a) \\ 2(\bar{q}_{a,0}\bar{x}_a + \bar{q}_{a,z}\bar{y}_a - \bar{q}_{a,y}\bar{z}_a) & -2(\bar{q}_{a,x}\bar{x}_a + \bar{q}_{a,y}\bar{y}_a + \bar{q}_{a,y}\bar{z}_a) \\ 2\bar{q}_{a,y} & 2\bar{q}_{a,z} \\ 2\bar{q}_{a,z} & -2\bar{q}_{a,y} \\ -2\bar{q}_{a,0} & 2\bar{q}_{a,x} \\ -2\bar{q}_{a,x} & -2\bar{q}_{a,0} \end{matrix}\right)$$

$$
\boldsymbol{H}_b = \begin{pmatrix}
1 & 0 & 0 & 2(\bar{q}_{b,y}\bar{z}_b - \bar{q}_{b,0}\bar{x}_b - \bar{q}_{b,z}\bar{y}_b) & -2(\bar{q}_{b,x}\bar{x}_b + \bar{q}_{b,y}\bar{y}_b + \bar{q}_{b,z}\bar{z}_b) & 2(\bar{q}_{b,x}\bar{y}_b - \bar{q}_{b,y}\bar{x}_b - \bar{q}_{b,0}\bar{z}_b) & 2(\bar{q}_{b,0}\bar{y}_b - \bar{q}_{b,z}\bar{x}_b + \bar{q}_{b,x}\bar{z}_b) \\
0 & 1 & 0 & 2(\bar{q}_{b,x}\bar{z}_b - \bar{q}_{b,z}\bar{x}_b - \bar{q}_{b,0}\bar{y}_b) & 2(\bar{q}_{b,y}\bar{x}_b - \bar{q}_{b,x}\bar{y}_b + \bar{q}_{b,0}\bar{z}_b) & -2(\bar{q}_{b,x}\bar{x}_b + \bar{q}_{b,y}\bar{y}_b + \bar{q}_{b,z}\bar{z}_b) & 2(\bar{q}_{b,y}\bar{z}_b - \bar{q}_{b,0}\bar{x}_b - \bar{q}_{b,z}\bar{y}_b) \\
0 & 0 & 1 & 2(\bar{q}_{b,x}\bar{y}_b - \bar{q}_{b,y}\bar{x}_b - \bar{q}_{b,0}\bar{z}_b) & 2(\bar{q}_{b,z}\bar{x}_b - \bar{q}_{b,0}\bar{y}_b - \bar{q}_{b,x}\bar{z}_b) & 2(\bar{q}_{b,0}\bar{x}_b + \bar{q}_{b,z}\bar{y}_b - \bar{q}_{b,y}\bar{z}_b) & -2(\bar{q}_{b,x}\bar{x}_b + \bar{q}_{b,y}\bar{y}_b + \bar{q}_{b,y}\bar{z}_b) \\
0 & 0 & 0 & 2\bar{q}_{b,0} & 2\bar{q}_{b,x} & 2\bar{q}_{b,y} & 2\bar{q}_{b,z} \\
0 & 0 & 0 & 2\bar{q}_{b,x} & -2\bar{q}_{b,0} & 2\bar{q}_{b,z} & -2\bar{q}_{b,y} \\
0 & 0 & 0 & 2\bar{q}_{b,y} & -2\bar{q}_{b,z} & -2\bar{q}_{b,0} & 2\bar{q}_{b,x} \\
0 & 0 & 0 & 2\bar{q}_{b,z} & 2\bar{q}_{b,y} & -2\bar{q}_{b,x} & -2\bar{q}_{b,0}
\end{pmatrix}
$$

To calculate the pose difference estimation $\bar{\mathbf{D}}$ and the covariance $\boldsymbol{C}_D$, we compute the $3m \times 7$ matrix $\mathbf{M}$ and the $3m$ vector $\mathbf{Z}$. These are given now as follows:

$$
\mathbf{M} = \begin{pmatrix}
1 & 0 & 0 & x_1 & 0 & -z_1 & y_1 \\
0 & 1 & 0 & y_1 & z_1 & 0 & -x_1 \\
0 & 0 & 1 & z_1 & -y_1 & x_1 & 0 \\
\vdots & \vdots & \vdots & \vdots & \vdots & \vdots & \\
1 & 0 & 0 & x_m & 0 & -z_m & y_m \\
0 & 1 & 0 & y_m & z_m & 0 & -x_m \\
0 & 0 & 1 & z_m & -y_m & x_m & 0
\end{pmatrix}
$$

$$
\mathbf{Z} = \begin{pmatrix}
\bar{\boldsymbol{V}}_a \oplus \boldsymbol{u}_1^a - \bar{\boldsymbol{V}}_b \oplus \boldsymbol{u}_1^b \\
\bar{\boldsymbol{V}}_a \oplus \boldsymbol{u}_2^a - \bar{\boldsymbol{V}}_b \oplus \boldsymbol{u}_2^b \\
\vdots \\
\bar{\boldsymbol{V}}_a \oplus \boldsymbol{u}_m^a - \bar{\boldsymbol{V}}_b \oplus \boldsymbol{u}_m^b
\end{pmatrix}.
$$

Theorem 10. *The covariance matrix* $\boldsymbol{C}_D = s^2(\mathbf{M}^T\mathbf{M})^{-1}$ *is positive definite.*

Proof. To show that the covariance $\boldsymbol{C}_D = s^2(\mathbf{M}^T\mathbf{M})^{-1}$ is positive definite we analyze $\mathbf{M}^T\mathbf{M}$ again.

$$\mathbf{M}^T\mathbf{M} = \sum_{k=1}^{m} \begin{pmatrix} 1 & 0 & 0 & x_k & 0 & -z_k & y_k \\ 0 & 1 & 0 & y_k & z_k & 0 & -x_k \\ 0 & 0 & 1 & z_k & -y_k & x_k & 0 \\ x_k & y_k & z_k & x_k^2 + y_k^2 + z_k^2 & 0 & 0 & 0 \\ 0 & z_k & -y_k & 0 & y_k^2 + z_k^2 & -x_k y_k & -x_k z_k \\ -z_k & 0 & x_k & 0 & -x_k y_k & x_k^2 + z_k^2 & -y_k z_k \\ y_k & -x_k & 0 & 0 & -x_k z_k & -y_k z_k & x_k^2 + y_k^2 \end{pmatrix}.$$

Using the definition of positive definite for every real vector $\boldsymbol{a} = (a_1, a_2, a_3, a_4, a_5, a_6, a_7)^T \neq 0$ we have:

$$\begin{aligned} \boldsymbol{a}^T\mathbf{M}^T\mathbf{M}\boldsymbol{a} = \sum_{k=1}^{m} & (a_1 + a_4 x_k + a_7 y_k - a_6 z_k)^2 \\ & + (a_2 - a_7 x_k + a_4 y_k + a_5 z_k)^2 \\ & + (a_3 + a_6 x_k - a_5 y_k - a_4 z_k)^2. \end{aligned}$$

The above equation is zero, if there exists an $(\boldsymbol{A}, \boldsymbol{B}) \neq \mathbf{0}$ such that for all $\boldsymbol{u}_k$ the following equation is true:

$$\boldsymbol{A}\boldsymbol{u}_k = \boldsymbol{B}$$

$$\begin{pmatrix} a_4 & a_7 & -a_6 \\ -a_7 & a_4 & a_5 \\ a_6 & -a_5 & a_4 \end{pmatrix} \cdot \begin{pmatrix} x_k \\ y_k \\ z_k \end{pmatrix} = \begin{pmatrix} -a_1 \\ -a_2 \\ -a_3 \end{pmatrix}.$$

If we set $\boldsymbol{A} = \mathbf{0}$ then we have $\boldsymbol{B} = \mathbf{0}$. There exist elements a_4, a_5, a_6 and a_7 that are not equal to zero. The matrix $\boldsymbol{A}$ has rank zero, if its determinant is zero.

$$\begin{aligned} \mathrm{Det}(\boldsymbol{A}) &= 0 \\ &= a_4^3 + a_4 a_5^2 + a_4 a_6^2 + a_4 a_7^2 \\ &= a_4(a_4^2 + a_5^2 + a_6^2 + a_7^2). \end{aligned}$$

This is only the case with $\boldsymbol{A} \neq 0$ if $a_4 = 0$. Therefore, $\boldsymbol{A}$ has the follwoing form:

$$\boldsymbol{A} = \begin{pmatrix} 0 & a_7 & -a_6 \\ -a_7 & 0 & a_5 \\ a_6 & -a_5 & 0 \end{pmatrix}.$$

Consequently, $\boldsymbol{A}$ where $a_4 = 0$ has the minimal rank 2 and the matrix $\boldsymbol{A}$ where $a_4 \neq 0$ has always full rank. This is the case, if the corresponding point pairs are not co-linear. Then $\boldsymbol{C}_D$ is positive definite. □

To use the quaternion based solution instead of the Euler one in Algorithm 5.1 we have to use the quaternion pose representation and coveriance estimation. In addition we need to normalize the quaternion after transforming the solution. This ensures that the quaternion represents a valid rotation.

5.3.2 GraphSLAM Using Helical Motion

A linearization of the transformation $(\boldsymbol{R}, \boldsymbol{t})$ was discussed in Section 4.1.2. The following subsection gives a simple extension such that it fits our GraphSLAM framework for simultaneous registration. The method is also globally optimal. It was developed by Pottman et al. [98].

First of all, we have to set up an error function based on scan matching, that has to be minimized. Here we sum similarly to Eq. (4.30) over the point-to-point distances for the link $(j \to k)$. For enhancing readability, we denote points belonging to scan j using $\boldsymbol{x}$ and points of scan k with $\boldsymbol{y}$:

$$E = \sum_{j \to k} \sum_{i=1}^{N} Q_{j,k}(\boldsymbol{x}_i, \boldsymbol{y}_i),$$

where $Q_{j,k}$ is defined as follows:

$$Q_{j,k}(\boldsymbol{x}_i, \boldsymbol{y}_i) = (\boldsymbol{x}_i + \boldsymbol{v}_{j->k}(\boldsymbol{x}_i) - \boldsymbol{y}_i) \tag{5.14}$$
$$= (\boldsymbol{x}_i - \boldsymbol{y}_i + (\bar{\boldsymbol{c}}_j + \boldsymbol{c}_j \times \boldsymbol{x}_i) - (\bar{\boldsymbol{c}}_k + \boldsymbol{c}_k \times \boldsymbol{x}_i))^2 \tag{5.15}$$

The helical motion $\boldsymbol{v}_{j->k}(\boldsymbol{x}_i)$ denotes the relative motion of the jth 3D scan with respect to the kth 3D scan. The equality between Eq. (5.14) and (5.15) holds, since this relative motion is expressible as:

$$\boldsymbol{v}_{j->k}(\boldsymbol{x}_i) = \boldsymbol{v}_{j->0}(\boldsymbol{x}_i) - \boldsymbol{v}_{k->0}(\boldsymbol{x}_i)$$

First of all, we use the terms $\boldsymbol{C}_j$ and $\boldsymbol{C}_k$ in order to simplify the following calculations. For Eq. (5.15) we have

$$
\begin{aligned}
&:= (\boldsymbol{x}_i - \boldsymbol{y}_i + \boldsymbol{C}_j - \boldsymbol{C}_k))^2 \\
&= \ (\boldsymbol{x}_i - \boldsymbol{y}_i)^2 + (\boldsymbol{C}_j - \boldsymbol{C}_k)^2 + 2\left((\boldsymbol{x}_i - \boldsymbol{y}_i)\ \boldsymbol{C}_j + (\boldsymbol{y}_i - \boldsymbol{x}_i)\ \boldsymbol{C}_k\right)
\end{aligned}
$$

and for the global error function

$$
\begin{aligned}
E = \sum_{j \to k} \sum_{i}^{N} (\boldsymbol{C}_j - \boldsymbol{C}_k)^2 \\
+ \sum_{i}^{N} 2((\boldsymbol{x}_i - \boldsymbol{y}_i)\ \boldsymbol{C}_j + (\boldsymbol{y}_i - \boldsymbol{x}_i)\ \boldsymbol{C}_k) + \sum_{i}^{N} (\boldsymbol{x}_i - \boldsymbol{y}_i)^2
\end{aligned}
$$

We transform the above equation in a quadratatic form, i.e.,

$$
E = \mathbf{C}^T \mathbf{B}\, \mathbf{C} + 2\, \mathbf{A}\, \mathbf{C} + \sum_{j \to k} \sum_{i}^{N} (\boldsymbol{x}_i - \boldsymbol{y}_i)^2 \tag{5.16}
$$

Then we find the optimal solution by setting the derivative of (5.16) to zero and solve for $\mathbf{C}$, that is the concatenation of $(\boldsymbol{c}_1, \bar{\boldsymbol{c}}_1, \boldsymbol{c}_2, \bar{\boldsymbol{c}}_2, \ldots, \boldsymbol{c}_n, \bar{\boldsymbol{c}}_n)$. Thus this concatenation contains the transformations for all 3D scans. For doing so, the single terms have to be written as:

$$
\sum_{j \to k} \sum_{i}^{N} 2((\boldsymbol{x}_i - \boldsymbol{y}_i)\ \boldsymbol{C}_j + (\boldsymbol{y}_i - \boldsymbol{x}_i)\ \boldsymbol{C}_k) = 2\, \mathbf{A}\, \mathbf{C}
$$

$$
\sum_{j \to k} \sum_{i}^{N} (\boldsymbol{C}_j - \boldsymbol{C}_k)^2 = \mathbf{C}^T \mathbf{B}\, \mathbf{C}
$$

Deriving A. To compute the entries of the matrix $\mathbf{A}$ we proceed as follows:

$$
\begin{aligned}
2\, \mathbf{A}\ \mathbf{C} &= 2 \sum_{j \to k} \sum_{i} ((\boldsymbol{x}_i - \boldsymbol{y}_i)\ \boldsymbol{C}_j + (\boldsymbol{y}_i - \boldsymbol{x}_i)\ \boldsymbol{C}_k) \\
&= 2 \sum_{j \to k} \sum_{i} \begin{pmatrix} \boldsymbol{x}_i - \boldsymbol{y}_i \\ \boldsymbol{y}_i - \boldsymbol{x}_i \end{pmatrix} \begin{pmatrix} \boldsymbol{C}_j \\ \boldsymbol{C}_k \end{pmatrix} \\
&= 2 \sum_{j \to k} \sum_{i} \begin{pmatrix} \boldsymbol{x}_i - \boldsymbol{y}_i \\ \boldsymbol{y}_i - \boldsymbol{x}_i \end{pmatrix} \begin{pmatrix} \bar{\mathbf{c}}_j \\ \bar{\mathbf{c}}_k \end{pmatrix} + \begin{pmatrix} \boldsymbol{x}_i - \boldsymbol{y}_i \\ \boldsymbol{y}_i - \boldsymbol{x}_i \end{pmatrix} \begin{pmatrix} \mathbf{c}_j \times \boldsymbol{x}_i \\ \mathbf{c}_k \times \boldsymbol{x}_i \end{pmatrix}
\end{aligned}
$$

using the equivalence of the parallelepipedal product:

$$=2\sum_{j\to k}\sum_i \begin{pmatrix} \boldsymbol{x}_i - \boldsymbol{y}_i \\ \boldsymbol{y}_i - \boldsymbol{x}_i \end{pmatrix} \begin{pmatrix} \bar{\boldsymbol{c}}_j \\ \bar{\boldsymbol{c}}_k \end{pmatrix} + \begin{pmatrix} \boldsymbol{x}_i \times (\boldsymbol{x}_i - \boldsymbol{y}_i) \\ \boldsymbol{x}_i \times (\boldsymbol{y}_i - \boldsymbol{x}_i) \end{pmatrix} \begin{pmatrix} \boldsymbol{c}_j \\ \boldsymbol{c}_k \end{pmatrix}$$

$$=2\sum_{j\to k}\sum_i \begin{pmatrix} \boldsymbol{x}_i \times (\boldsymbol{x}_i - \boldsymbol{y}_i) \\ \boldsymbol{x}_i - \boldsymbol{y}_i \\ \boldsymbol{x}_i \times (\boldsymbol{y}_i - \boldsymbol{x}_i) \\ \boldsymbol{y}_i - \boldsymbol{x}_i \end{pmatrix} \begin{pmatrix} \boldsymbol{c}_j \\ \bar{\boldsymbol{c}}_j \\ \boldsymbol{c}_k \\ \bar{\boldsymbol{c}}_k \end{pmatrix}$$

Thus $\mathbf{A} = (\boldsymbol{A}_1, \boldsymbol{A}_{\bar{1}}, \ldots, \boldsymbol{A}_{\bar{N}})$ that is the concatenation of $\boldsymbol{A}_j$ und $\boldsymbol{A}_{\bar{j}}$ is formed using the closest point pairs $\boldsymbol{x}_{i,j}$ und $\boldsymbol{y}_{i,j}$ as given below. The additional index j denotes the associated 3D scan.

$$\boldsymbol{A}_j = \sum_{j\to k}\sum_i \boldsymbol{x}_{i,j} \times (\boldsymbol{x}_{i,j} - \boldsymbol{y}_{i,k}) + \sum_{k\to j}\sum_i \boldsymbol{x}_{i,k} \times (\boldsymbol{y}_{i,j} - \boldsymbol{x}_{i,k})$$

$$\boldsymbol{A}_{\bar{j}} = \sum_{j\to k}\sum_i (\boldsymbol{x}_{i,j} - \boldsymbol{y}_{i,k}) + \sum_{k\to j}\sum_i (\boldsymbol{x}_{i,j} - \boldsymbol{y}_{i,k})$$

Expressed understandable: For every link $j \to k$ between two 3D scans one has to add for $\boldsymbol{A}_j$ the terms $\sum_i \boldsymbol{x}_{i,j} \times (\boldsymbol{x}_{i,j} - \boldsymbol{y}_{i,j})$ and for $\boldsymbol{A}_k$ the terms $\sum_i \boldsymbol{x}_{i,k} \times (\boldsymbol{y}_{i,j} - \boldsymbol{x}_{i,k})$. In the same way $\boldsymbol{A}_{\bar{j}}$ and $\boldsymbol{A}_{\bar{k}}$ are computed.

Deriving B. The matrix **B** has the form:

$$\mathbf{B} = \begin{pmatrix} \boldsymbol{B}_{1,1} & \boldsymbol{B}_{1,\bar{1}} & \boldsymbol{B}_{1,2} & \boldsymbol{B}_{1,\bar{2}} & \cdots & \boldsymbol{B}_{1,\bar{N}} \\ \boldsymbol{B}_{\bar{1},1} & \boldsymbol{B}_{\bar{1},\bar{1}} & & & & \\ \boldsymbol{B}_{2,1} & & \boldsymbol{B}_{2,2} & & & \\ \vdots & & & \boldsymbol{B}_{\bar{2},\bar{2}} & & \\ & & & & \ddots & \\ \boldsymbol{B}_{\bar{N},1} & & & & & \boldsymbol{B}_{\bar{N},\bar{N}} \end{pmatrix}.$$

Every submatrix is a 3×3 matrix that is filled using quotient comparison.

$$\begin{aligned} \mathbf{C}^T\,\mathbf{B}\,\mathbf{C} &= \sum_{j\to k}\sum_i (\boldsymbol{C}_j - \boldsymbol{C}_k)^2 \\ &= \sum_{j\to k}\sum_i \boldsymbol{C}_j^2 + \boldsymbol{C}_k^2 - 2\boldsymbol{C}_j\,\boldsymbol{C}_k \end{aligned}$$

The term $\boldsymbol{C}_j^2 + \boldsymbol{C}_k^2$ forms the block diagonal of the Matrix $\mathbf{B}$, while $-2\,\boldsymbol{C}_j\,\boldsymbol{C}_k$ fills the remaining parts:

$$\begin{aligned}\boldsymbol{C}_j^2 + \boldsymbol{C}_k^2 = \bar{\boldsymbol{c}}_j^2 + (\mathbf{c}_j \times \boldsymbol{x}_i)^2 + \bar{\mathbf{c}}_k^2 + (\mathbf{c}_k \times \boldsymbol{x}_i)^2 \\ + 2\bar{\mathbf{c}}_j\,(\mathbf{c}_j \times \boldsymbol{x}_i) + 2\bar{\mathbf{c}}_k\,(\mathbf{c}_k \times \boldsymbol{x}_i)\end{aligned}$$

Using the Lagrange identity we derive:

$$\begin{aligned}= \bar{\mathbf{c}}_j + \boldsymbol{x}_i^2\,\mathbf{c}_j^2 - (\mathbf{c}_j\,\boldsymbol{x}_i)^2 + \bar{\mathbf{c}}_k^2 + \boldsymbol{x}_i^2\,\mathbf{c}_k^2 - (\mathbf{c}_k\,\boldsymbol{x}_i)^2 \\ + 2\bar{\mathbf{c}}_j\,(\mathbf{c}_j \times \boldsymbol{x}_i) + 2\bar{\mathbf{c}}_k\,(\mathbf{c}_k \times \boldsymbol{x}_i)\end{aligned}$$

The submatrix $\boldsymbol{B}_{\bar{j},\bar{j}}$ is therefore given by

$$\boldsymbol{B}_{\bar{j},\bar{j}} = \sum_{\substack{j\to k) \\ k\to j}} \sum_i \mathbb{1}_{3\times 3}$$

The submatrix $\boldsymbol{B}_{j,j}$ is the results of $\boldsymbol{x}_i^2\,\mathbf{c}_j^2 - (\mathbf{c}_j\,\boldsymbol{x}_i)^2$ und $\boldsymbol{x}_i^2\,\mathbf{c}_k^2 - (\mathbf{c}_k\,\boldsymbol{x}_i)^2$

$$\boldsymbol{B}_{j,j} = \sum_{\substack{j\to k \\ k\to j}} \sum_i^N \begin{pmatrix} \boldsymbol{x}_i^2 - x_{x,i}^2 & -x_{x,i}x_{y,i} & -x_{x,i}x_{z,i} \\ -x_{x,i}x_{y,i} & \boldsymbol{x}_i^2 - x_{y,i}^2 & -x_{y,i}x_{z,i} \\ -x_{x,i}x_{z,i} & -x_{y,i}x_{z,i} & \boldsymbol{x}_i^2 - x_{z,i}^2 \end{pmatrix}.$$

The terms $2\,\bar{\mathbf{c}}_j\,(\mathbf{c}_j \times \boldsymbol{x}_i)$ und $2\bar{\mathbf{c}}_k\,(\mathbf{c}_k \times \boldsymbol{x}_i)$ lead to the submatrices $\boldsymbol{B}_{\bar{j},j}$ and $\boldsymbol{B}_{j,\bar{j}}$:

$$\boldsymbol{B}_{j,\bar{j}} = -\boldsymbol{B}_{\bar{j},j} = \sum_{\substack{(j,k) \\ (k,j)}} \sum_i \begin{pmatrix} 0 & x_{z,i}^2 & -x_{y,i}^2 \\ -x_{z,i}^2 & 0 & x_{x,i}^2 \\ x_{y,i}^2 & -x_{x,i}^2 & 0 \end{pmatrix}.$$

The remaining parts of $\mathbf{B}$ are given by $-2\,\boldsymbol{C}_j\,\boldsymbol{M}_k$

$$\begin{aligned}-2\boldsymbol{C}_j\,\boldsymbol{C}_k = -2(&\bar{\mathbf{c}}_j\,\bar{\mathbf{c}}_k \\ &+ (\mathbf{c}_j \times \boldsymbol{x}_i)\,(\mathbf{c}_k \times \boldsymbol{x}_i) \\ &+ \bar{\mathbf{c}}_j\,(\mathbf{c}_k \times \boldsymbol{x}_i) + \bar{\mathbf{c}}_k\,(\mathbf{c}_j \times \boldsymbol{x}_i)).\end{aligned}$$

Again we use the Lagrange identity to calculate:

$$
\begin{aligned}
&= -2(\bar{\boldsymbol{c}}_j\ \bar{\boldsymbol{c}}_k \\
&\qquad + (\boldsymbol{c}_j\ \boldsymbol{c}_k)\ \boldsymbol{x}_i^2 - (\boldsymbol{c}_j\ \boldsymbol{x}_i)\ (\boldsymbol{c}_k\ \boldsymbol{x}_i) \\
&\qquad + \bar{\boldsymbol{c}}_j\ (\boldsymbol{c}_k \times \boldsymbol{x}_i) + \bar{\boldsymbol{c}}_k\ (\boldsymbol{c}_j \times \boldsymbol{x}_i))
\end{aligned}
$$

Now $\boldsymbol{B}_{\bar{j},\bar{k}}$ where $j \neq k$ is given as

$$
\boldsymbol{B}_{\bar{j},\bar{k}} = \boldsymbol{B}_{\bar{k},\bar{j}} = -\sum_{j \to k} \sum_i \mathbb{1}_{3\times 3}.
$$

Similarily the submatrix $\boldsymbol{B}_{j,k}$ where $j \neq k$ is computed:

$$
\boldsymbol{B}_{j,k} = -\sum_{j \to k} \sum_i \begin{pmatrix} \boldsymbol{x}_i^2 - x_{x,i}^2 & -x_{x,i}x_{y,i} & -x_{x,i}x_{z,i} \\ -x_{x,i}x_{y,i} & \boldsymbol{x}_i^2 - x_{y,i}^2 & -x_{y,i}x_{z,i} \\ -x_{x,i}x_{z,i} & -x_{y,i}x_{z,i} & \boldsymbol{x}_i^2 - x_{z,i}^2 \end{pmatrix}.
$$

In the same way $\boldsymbol{B}_{\bar{j},k}$ as well as $\boldsymbol{B}_{j,\bar{k}}$ can be computed:

$$
\boldsymbol{B}_{j,\bar{k}} = \boldsymbol{B}_{\bar{j},k} = \sum_{(j,k)} \sum_i \begin{pmatrix} 0 & -x_{z,i}^2 & x_{y,i}^2 \\ x_{z,i}^2 & 0 & -x_{x,i}^2 \\ -x_{y,i}^2 & x_{x,i}^2 & 0 \end{pmatrix}.
$$

Putting it together. To solve GraphSLAM, we have to fix the first scan, which implies that $(\boldsymbol{c}_1, \bar{\boldsymbol{c}}_1)$ is set to $(\mathbf{0}, \mathbf{0})$. Now, we are using a simpler notation and summarize the previous calculations. The $1 \times 6(N-1)$ vector $\mathbf{A} = (\boldsymbol{A}_2, \boldsymbol{A}_3, \ldots, \boldsymbol{A}_N)$ consists of 1×6 vectors $\boldsymbol{A}_j$, that have the following form

$$
\boldsymbol{A}_j = \sum_{j \to k} \sum_i \begin{pmatrix} \boldsymbol{x}_{i,j} \times (\boldsymbol{x}_{i,j} - \boldsymbol{y}_{i,k}) \\ (\boldsymbol{x}_{i,j} - \boldsymbol{y}_{i,k}) \end{pmatrix} + \sum_{k \to j} \sum_i \begin{pmatrix} \boldsymbol{x}_{i,k} \times (\boldsymbol{y}_{i,j} - \boldsymbol{x}_{i,k}) \\ (\boldsymbol{y}_{i,j} - \boldsymbol{x}_{i,k}) \end{pmatrix}
$$

The $6(N-1) \times 6(N-1)$ matrix $\mathbf{B}$ is now expressed in a more compact way, i.e., using 6×6 submatrices $\boldsymbol{B}_{j,k}$:

$$
\boldsymbol{B} = \begin{pmatrix} \boldsymbol{B}_{2,2} & \boldsymbol{B}_{2,3} & \cdots & \boldsymbol{B}_{2,N} \\ \boldsymbol{B}_{3,2} & \boldsymbol{B}_{3,3} & & \\ \vdots & & \ddots & \vdots \\ \boldsymbol{B}_{N,2} & & \cdots & \boldsymbol{B}_{N,N} \end{pmatrix}.
$$

If $j = k$:

$$
\boldsymbol{B}_{j,j} = \sum_{\substack{j \to k \\ k \to j}} \sum_i \begin{pmatrix}
\boldsymbol{x}_i^2 - x_{x,i}^2 & -x_{x,i}x_{y,i} & -x_{x,i}x_{z,i} & 0 & -x_{z,i} & x_{y,i} \\
-x_{x,i}x_{y,i} & \boldsymbol{x}_i^2 - x_{y,i}^2 & -x_{y,i}x_{z,i} & x_{z,i} & 0 & -x_{x,i} \\
-x_{x,i}x_{z,i} & -x_{y,i}x_{z,i} & \boldsymbol{x}_i^2 - x_{z,i}^2 & -x_{y,i} & x_{x,i} & 0 \\
0 & x_{z,i} & -x_{y,i} & 1 & 0 & 0 \\
-x_{z,i} & 0 & x_{x,i} & 0 & 1 & 0 \\
x_{y,i} & -x_{x,i} & 0 & 0 & 0 & 1
\end{pmatrix}.
$$

If $j \neq k$:

$$
\boldsymbol{B}_{j,k} = \sum_{(j,k)} \sum_i \begin{pmatrix}
x_{x,i}^2 - \boldsymbol{x}_i^2 & x_{x,i}x_{y,i} & x_{x,i}x_{z,i} & 0 & x_{z,i} & -x_{y,i} \\
x_{x,i}x_{y,i} & x_{y,i}^2 - \boldsymbol{x}_i^2 & x_{y,i}x_{z,i} & -x_{z,i} & 0 & x_{x,i} \\
x_{x,i}x_{z,i} & x_{y,i}x_{z,i} & x_{z,i}^2 - \boldsymbol{x}_i^2 & x_{y,i} & -x_{x,i} & 0 \\
0 & -x_{z,i} & x_{y,i} & -1 & 0 & 0 \\
x_{z,i} & 0 & -x_{x,i} & 0 & -1 & 0 \\
-x_{y,i} & x_{x,i} & 0 & 0 & 0 & -1
\end{pmatrix}.
$$

5.3.3 GraphSLAM Using Linearization of a Rotation

Similarily to Section 4.1.3 where we have extended the results of Section 4.1.2, we now look at the previous solution again an apply a different liniearization scheme. We will linearize the rotation as given in Eq. (4.35) and (4.36) and apply the centroid trick again, this time in gloabal calculations.

The error $E(j, k)$ between the two scans j, k is given by the point correspondencies $(\boldsymbol{x}_i, \boldsymbol{y}_i)$, where $\boldsymbol{x}_i$ is a 3D point in scan j and $\boldsymbol{y}_i$ in k:

$$E(j,k) = \sum_i ||\boldsymbol{R}_j \boldsymbol{x}_i + \boldsymbol{t}_j - (\boldsymbol{R}_k \boldsymbol{y}_i + \boldsymbol{t}_k)||^2$$

Using the centroids $\boldsymbol{c}_y$ and $\boldsymbol{c}_x$, that are given by $\boldsymbol{x}'_i = \boldsymbol{x}_i - \boldsymbol{c}_x$, $\boldsymbol{y}'_i = \boldsymbol{y}_i - \boldsymbol{c}_y$, the above error function can be transformed as follows:

$$\begin{aligned} E(j,k) &= \sum_i ||\boldsymbol{R}_j \boldsymbol{x}'_i + \boldsymbol{R}_j \boldsymbol{c}_x + \boldsymbol{t}_j - (\boldsymbol{R}_k \boldsymbol{d}'_i + \boldsymbol{R}_k \boldsymbol{c}_y + \boldsymbol{t}_k)||^2 \\ &= \sum_i ||\boldsymbol{R}_j \boldsymbol{x}'_i - \boldsymbol{R}_k \boldsymbol{y}'_i - (\boldsymbol{t}_k - \boldsymbol{t}_j + \boldsymbol{R}_k \boldsymbol{c}_y - \boldsymbol{R}_j \boldsymbol{c}_x)||^2 \\ &= \sum_i ||\boldsymbol{R}_j \boldsymbol{x}'_i - \boldsymbol{R}_k \boldsymbol{y}'_i||^2 \\ &\quad - 2 \sum_i (\boldsymbol{t}_k - \boldsymbol{t}_j + \boldsymbol{R}_k \boldsymbol{c}_y - \boldsymbol{R}_j \boldsymbol{c}_x)(\boldsymbol{R}_j \boldsymbol{x}'_i - \boldsymbol{R}_k \boldsymbol{y}'_i) \\ &\quad + \sum_i ||\boldsymbol{t}_k - \boldsymbol{t}_j + \boldsymbol{R}_k \boldsymbol{c}_y - \boldsymbol{R}_j \boldsymbol{c}_x||^2 \end{aligned}$$

Similar to the argumentation on page 38 the middle term is zero, since the values refer to the centroid. As extention to the ICP approach the error over all relevant scan pairs is minimized

$$\begin{aligned} E &= \sum_{j \to k} E(j,k) \\ &= \sum_{j \to k} \sum_i ||\boldsymbol{R}_j \boldsymbol{x}'_i - \boldsymbol{R}_k \boldsymbol{y}'_i||^2 + \sum_{j \to k} \sum_i ||\boldsymbol{t}_k - \boldsymbol{t}_j + \boldsymbol{R}_k \boldsymbol{c}_y - \boldsymbol{R}_j \boldsymbol{c}_x||^2 \end{aligned}$$

Calculating the Optimal Rotation. To calculate the optimal rotation we minimize the first term using the linearization (4.36):

$$\begin{aligned} \boldsymbol{R}_j \boldsymbol{x}_i &= \begin{pmatrix} 0 & x_{z,i} & -x_{y,i} \\ -x_{z,i} & 0 & x_{x,i} \\ x_{y,i} & -x_{x,i} & 0 \end{pmatrix} \boldsymbol{X}_j + \boldsymbol{x}_i \\ &= \boldsymbol{M}_i \, \boldsymbol{X}_j + \boldsymbol{x}_i \end{aligned}$$

The rotation is the vector $\boldsymbol{X}_j = (\theta_z, \theta_y, \theta_x)^T$. To find the optimal rotation, we have to minimize the following sum:

$$F = \sum_{j \to k} \sum_i ||\boldsymbol{M}_i \, \boldsymbol{X}_j - \boldsymbol{D}_i \, \boldsymbol{X}_k - (\boldsymbol{x}_i - \boldsymbol{y}_i)||^2$$

respectively,

$$\begin{aligned} F = & \sum_{j \to k} \sum_i \left(\boldsymbol{M}_i \boldsymbol{X}_j - \boldsymbol{D}_i \boldsymbol{X}_k - (\boldsymbol{x}_i - \boldsymbol{y}_i) \right)^2 \\ = & \sum_{j \to k} \sum_i \left(\boldsymbol{M}_i \boldsymbol{X}_j - \boldsymbol{D}_i \boldsymbol{X}_k \right)^2 \\ & - 2 \left(\boldsymbol{M}_i \boldsymbol{X}_j - \boldsymbol{D}_i \boldsymbol{X}_i \right) \left((\boldsymbol{x}_i - \boldsymbol{y}_i) \right) \\ & + \left((\boldsymbol{x}_i - \boldsymbol{y}_i) \right)^2 \end{aligned}$$

The matrix $\boldsymbol{D}_i$ has the same form as the matrix $\boldsymbol{M}_i$. Using matrix notation F is represented as

$$F = \mathbf{X}\,\mathbf{B}\,\mathbf{X} + 2\,\mathbf{A}\,\mathbf{X} + \left((\boldsymbol{m}_k - \boldsymbol{d}_k) \right)^2$$

and therefore we vae to solve the linear system

$$\mathbf{B}\,\mathbf{X} + \mathbf{A} = 0$$

to find the optimal rotation. Now we have to find the matrix $\mathbf{B}$ and the vector $\mathbf{A}$. $\mathbf{B}$ is given by the terms:

$$\left(\boldsymbol{M}_i \boldsymbol{X}_j - \boldsymbol{D}_i \boldsymbol{X}_k \right)^2 = (\boldsymbol{M}_i \boldsymbol{X}_j)^2 + (\boldsymbol{D}_i \boldsymbol{X}_k)^2 - 2(\boldsymbol{M}_i \boldsymbol{X}_j)(\boldsymbol{D}_i \boldsymbol{X}_k)$$

Therefore, the entries of $\boldsymbol{B}$ are given by:

$$\boldsymbol{B}_{j,j} = \sum_{j \to k} \sum_i \boldsymbol{D}_i^T \boldsymbol{D}_i + \sum_{k \to j} \sum_i \boldsymbol{M}_i^T \boldsymbol{M}_i$$

If $j < k$

$$\boldsymbol{B}_{j,k} = - \sum_{j \to k} \sum_i \boldsymbol{M}_i^T \boldsymbol{D}_i$$

and if $j > k$

$$\boldsymbol{B}_{j,k} = - \sum_{j \to k} \sum_i \boldsymbol{D}_i^T \boldsymbol{M}_i.$$

$\mathbf{A}$ is formed by

$$-2 \left(\boldsymbol{M}_i \boldsymbol{X}_j - \boldsymbol{D}_i \boldsymbol{X}_k \right) \left(\boldsymbol{x}_i - \boldsymbol{y}_i \right)$$

such that $\boldsymbol{A}$ fills up as follows:

$$\boldsymbol{A}_j = \sum_{k\to j}\sum_i \begin{pmatrix} (x_{z,i}-y_{z,i})\,y_{y,i}-(x_{y,i}-y_{y,i})\,y_{z,i} \\ (x_{x,i}-y_{x,i})\,y_{z,i}-(x_{z,i}-y_{z,i})\,y_{x,i} \\ (x_{y,i}-y_{y,i})\,y_{x,i}-(x_{x,i}-y_{x,i})\,y_{y,i} \end{pmatrix}$$
$$-\sum_{j\to k}\sum_i \begin{pmatrix} (x_{z,i}-y_{z,i})\,x_{y,i}-(x_{y,i}-y_{y,i})\,x_{z,i} \\ (x_{x,i}-y_{x,i})\,x_{z,i}-(x_{z,i}-y_{z,i})\,x_{x,i} \\ (x_{y,i}-y_{y,i})\,x_{x,i}-(x_{x,i}-y_{x,i})\,x_{y,i} \end{pmatrix}$$

Again we notice the similarities to the results we obtained for the global helical motion.

Calculating the Optimal Translation. Using the optimal rotation the translation is calculated from

$$\sum_{j\to k}\sum_i ||\boldsymbol{t}_k-\boldsymbol{t}_j+\boldsymbol{R}_k\boldsymbol{c}_d-\boldsymbol{R}_j\boldsymbol{c}_m||^2 .$$

First of all, we replace $\boldsymbol{R}_k\boldsymbol{c}_y - \boldsymbol{R}_j\boldsymbol{c}_x$ with $\boldsymbol{R}$. Therefore, we minimize the follwoing sum:

$$F = \sum_{(j,k)}\sum_i (\boldsymbol{t}_k-\boldsymbol{t}_j)^2 - 2(\boldsymbol{t}_k-\boldsymbol{t}_j)\boldsymbol{R}+\boldsymbol{R}^2$$

In matrix notation we have

$$F = \mathbf{T}^T\,\mathbf{B}\,\mathbf{T} + 2\,\mathbf{A}\,\mathbf{T} + \mathbf{R}^2,$$

such that the linear system

$$\mathbf{B}\,\mathbf{T} + \mathbf{A} = 0$$

has to be solved. $\mathbf{B}$ is given by the terms

$$(\boldsymbol{t}_k-\boldsymbol{t}_j)^2 = \boldsymbol{t}_k^2+\boldsymbol{t}_j^2-2\boldsymbol{t}_k\boldsymbol{t}_j$$

and the submatrices of $\boldsymbol{B}$ have the following form

$$\boldsymbol{B}_{j,j} = \sum_{\substack{j\to k\\ k\to j}} \mathbb{1}_{3\times 3},$$

if $j<k$

$$\boldsymbol{B}_{j,k} = -\sum_{j\to k} \mathbb{1}_{3\times 3},$$

an if $j > k$

$$\boldsymbol{B}_{j,k} = -\sum_{j \to k} \mathbb{1}_{3\times 3}$$

Finally, **A** is given by:

$$\boldsymbol{A}_j = \sum_{j \to k} \boldsymbol{R}_j \boldsymbol{c}_x - \boldsymbol{R}_k \boldsymbol{c}_y - \sum_{k \to j} \boldsymbol{R}_j \boldsymbol{c}_x - \boldsymbol{R}_k \boldsymbol{c}_y$$

Chapter 6
Experiments and Results

6.1 3D Mapping with Kurt3D

6.1.1 The 3D Laser Range Finder

The 3D laser range finder is built on the basis of a SICK 2D range finder by extension with a mount and a small servomotor (Figure 6.1) [119]. The 2D laser range finder is attached to the mount at the center of rotation for achieving a controlled pitch motion with a standard servo.

Fig. 6.1 Kurt3D's 3D laser range finder. Its technical basis is a SICK 2D laser range finder (LMS-200).

The area of up to 180°(h) × 120°(v) is scanned with different horizontal (181, 361, 721) and vertical (128, 225, 420, 500) resolutions. A plane with 181 data points is scanned in 13 ms by the 2D laser range finder (rotating mirror device). Planes with more data points, e.g., 361, 721, duplicate or

A. Nüchter: 3D Robotic Map.: The Simultaneous Local. & Map. Prob., STAR 52, pp. 109–154.
springerlink.com

Fig. 6.2 Kurt3D in a natural environment. Top row: Lawn. Bottom left: Forest track. Bottom right: Pavement. Pictures taken at the Fraunhofer Campus Schloß Birlinghoven.

quadruplicate this time. Thus a scan with 181 × 256 data points needs 3.4 seconds. Scanning the environment with a mobile robot is done in a stop-scan-go fashion.

6.1.2 The Mobile Robot Kurt3D

The mobile robot Kurt3D in its outdoor version (Figure 6.2) is a mobile robot with a size of 45 cm (length) × 33 cm (width) × 29 cm (height) and a weight of 22.6 kg. Two 90 W motors are used to power the 6 skid-steered wheels, whereas the front and rear wheels have no tread pattern to enhance rotating. The core of the robot is a Pentium-Centrino-1400 with 768 MB RAM and Linux. An embedded 16-Bit CMOS microcontroller is used to control the motor.

6.1.3 Full 6D SLAM in a Planar Environment

We did apply the algorithms to a data set acquired at Schloss Dagstuhl. It contains 84 3D scans, each with 81225 (361×225), 3D data points, in a large, i.e., ~240 m, closed loop. The average distance between consecutive 3D scans was 2.5 m. Figure 6.3 shows the constructed 3D map.

We have also tested the algorithms on a 2D data set, computed from horizontal scan slices. This allows us to compare full 6D SLAM with planar 2D mapping. Figure 6.4 shows the map. There are noticeable errors in the 2D alignment. The 2D scans do not provide enough structure for correct alignment. Many approaches bypass this problem by scanning the

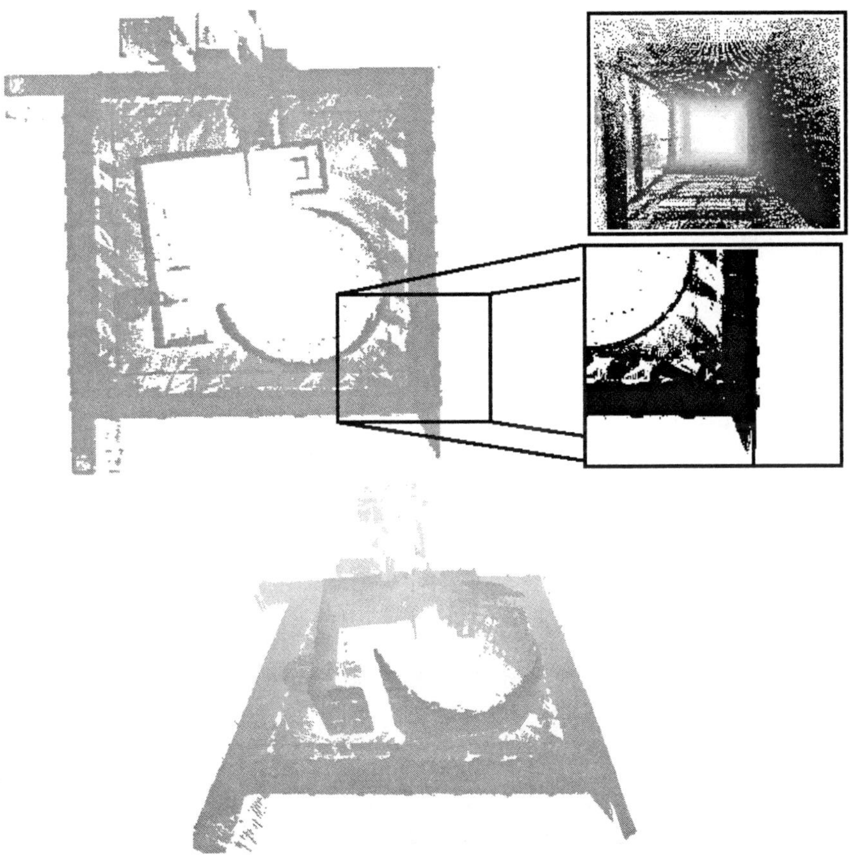

Fig. 6.3 3D digitalization of the International Conference and Research Center Schloss Dagstuhl. Top: 3D point cloud (top view). Bottom: 3D view.

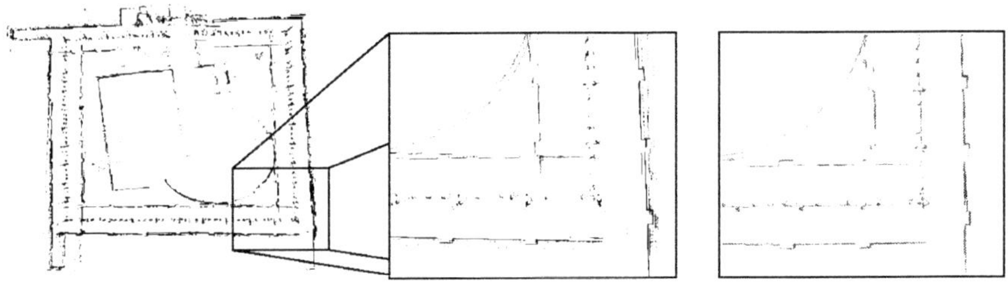

Fig. 6.4 2D digitalization of the environment with alignment problems at the wall on the right. Right: For comparison, the same closeup area of a horizontal slice from the generated 3D map (cf. Figure 6.3).

environment more often. However, the 3D data is much richer in information, therefore 3D scans taken at sparse discrete locations are matched correctly (cf. Figure 6.3).

The second experiment in a planar environment was performed at the University of Osnabrück. It features two loops. The robot's path, following the shape of an eight, is shown in the upper left corner of Figure 6.6. Figure 6.5 displays a top view of the floor. The left side shows the prearranged laser scans, the right shows the resulting alignment after correction with 900 iterations of Algorithm 5.1. A more detailed view of the map is presented in Figure 6.6. The initial error of up to 18 degrees is reduced significantly during global registration. The total error is distributed over all laser scans rather than to be summed up with each additional laser scan as is the case in iterative scan matching approaches.

6.1.4 Full 6D SLAM in an Indoor/Outdoor Environment

The proposed algorithms have been applied to a data set acquired at the robotic lab in Birlinghoven. 32 3D scans, each containing 302820 (721×420) range data points, were taken. The robot had to cope with a height difference between two buildings of 1.05 meter, covered, on the one hand, by a sloped driveway in open outdoor terrain, and, on the other hand, by a ramp of 12° inside the building. The 3D model was computed after acquiring all 3D scans.

Figure 4.1, 6.7 and 6.8 show rendered 3D scans. The latter figure presents the final model with the closed loop. Please refer to the website `http://kos.informatik.uni-osnabrueck.de/download/6Dpre/` for a computed animation and video through the scanned 3D scene.

In this data set, we analyzed the performance of the ICP scan matching. Details about this analysis can be found in Section 4.2 and in [86]. For ICP registration, the error tolerance for the initial estimation, i.e., the robot's self localization, is about 1meter in (x, y, z) position and about 15° in the orientation. These conditions can be easily met using a cheap odometer.

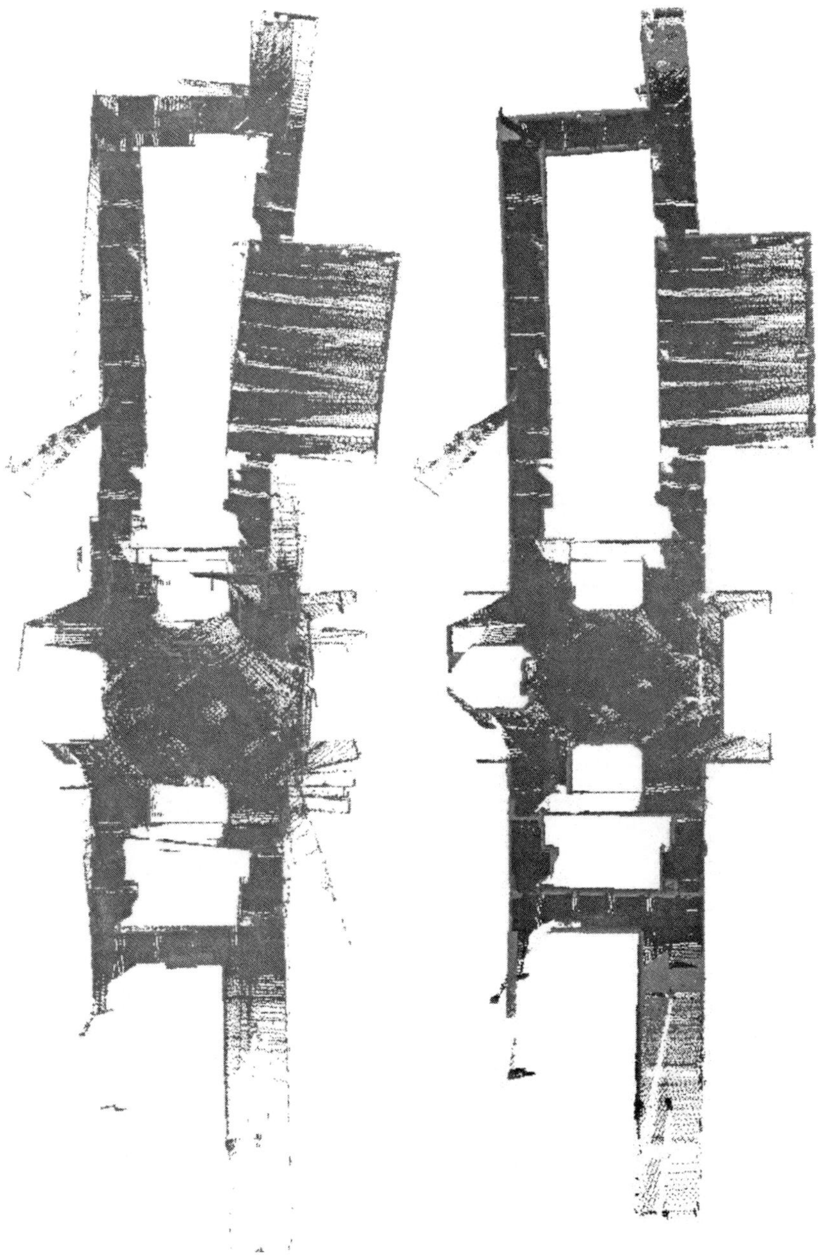

Fig. 6.5 Top view of 3D laser scans before (top) and after correction (bottom) with 900 iterations of Algorithm 5.1. The data result from 64 scans each containing 81225 (225 × 361) data points.

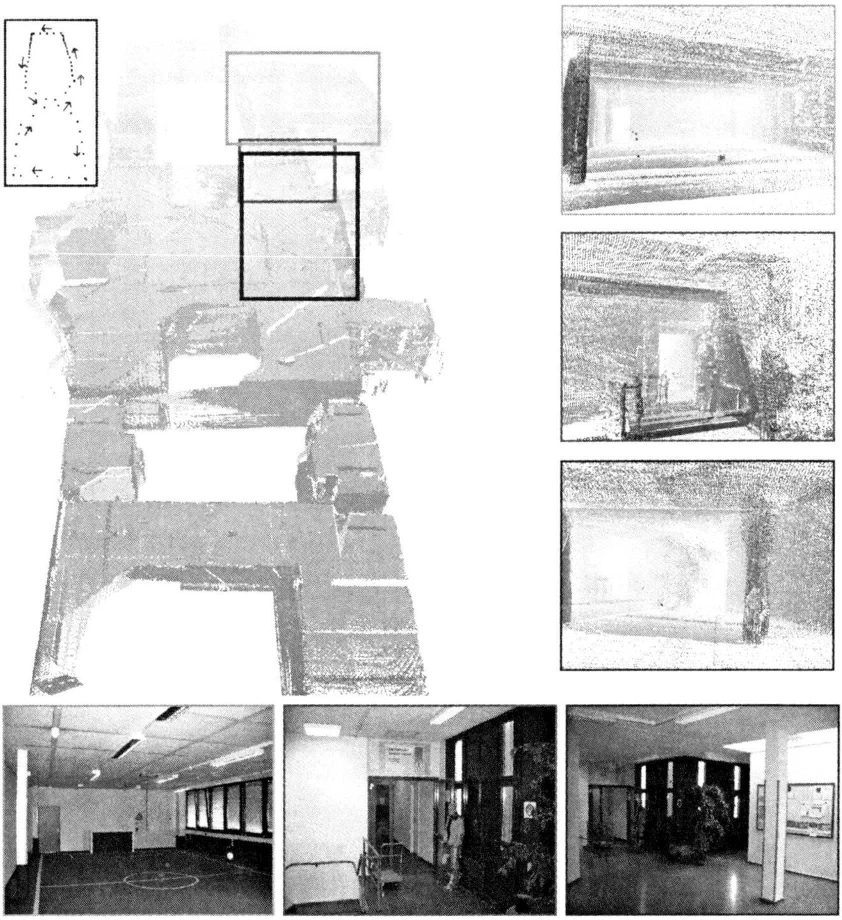

Fig. 6.6 Final 3D map. 3D points are classified and printed in different grey values. The closeups on the right correspond to the rectangles marked in the map. Top: Robot soccer field. Middle: Start and end of the robot path, showing a stair rail and a person standing in a doorway. Bottom: Part of the hall where the robot path intersects. The squares indicate the robot poses. In the top left corner, the complete robot path of approximately 127 meters is shown.

6.1.5 Full 6D SLAM in an Outdoor Environment

Single Loop Closing

The following experiment has been made at the Fraunhofer campus of Schloss Birlinghoven with Kurt3D. Figure 6.9 (left) shows the scan point model of the first scans in top view, based on odometry only. The first part of the robot's run, i.e., driving on asphalt, contains a systematic drift error, but driving on lawn shows more stochastic characteristics. The right part shows

Fig. 6.7 Left: Initial poses of 3D scans when closing the loop. Middle: Poses after detecting the loop and equally sharing the resulting error. Right: Final alignment after error diffusion with correct alignment of the edge structure at the ceiling.

the first 62 scans, covering a path length of about 240 m. The open loop is marked with a rectangle.

At that point, the loop is detected and closed. More 3D scans have then been acquired and added to the map. Figure 6.10 (left and right) shows the model with and without global relaxation to visualize its effects. The relaxation is able to align the scans correctly even without explicitly closing the loop. The best visible difference is marked by a rectangle. The final map in Figure 6.10 contains 77 3D scans, each consisting of approx. 100000 data points (361 × 275). Figure 6.11 shows two detailed views, before and after loop closing. The bottom part of Figure 6.10 displays an aerial view as ground truth for comparison. Table 6.1 compares distances measured in the photo and in the 3D scene, at corresponding points (taking roof overhangs into account). The lines in the photo have been measured in pixels, whereas real distances, i.e., the (x, z)-values of the points, have been used in the point model. Considering that pixel distances in mid-resolution non-calibrated aerial image induce some error in ground truth, the correspondences show that the point model at least approximates the reality quite well.

Mapping would fail without first extrapolating the odometry to generate a starting guess for ICP scan matching, since ICP would likely converge to

Table 6.1 Length ratio comparison of measured distances in the aerial photographs with distances in the point model as shown in Figure 6.10

1st line	2nd line	ratio in aerial views	ratio in point model	deviation
AB	BC	0.683	0.662	3.1%
AB	BD	0.645	0.670	3.8%
AC	CD	1.131	1.141	0.9%
CD	BD	1.088	1.082	0.5%

Fig. 6.8 The closed loop with a top viewing position and orthogonal projection. The distance d measured in the point cloud model is 2096 cm, measured by meter rule 2080 cm. The right part demonstrates the change in elevation. Top right: A ramp connecting two buildings is correctly modeled (height difference 1.05 m). The ramp connects the basement of the left building with the right building. Bottom right: Outdoor environment modeling of the downhill part.

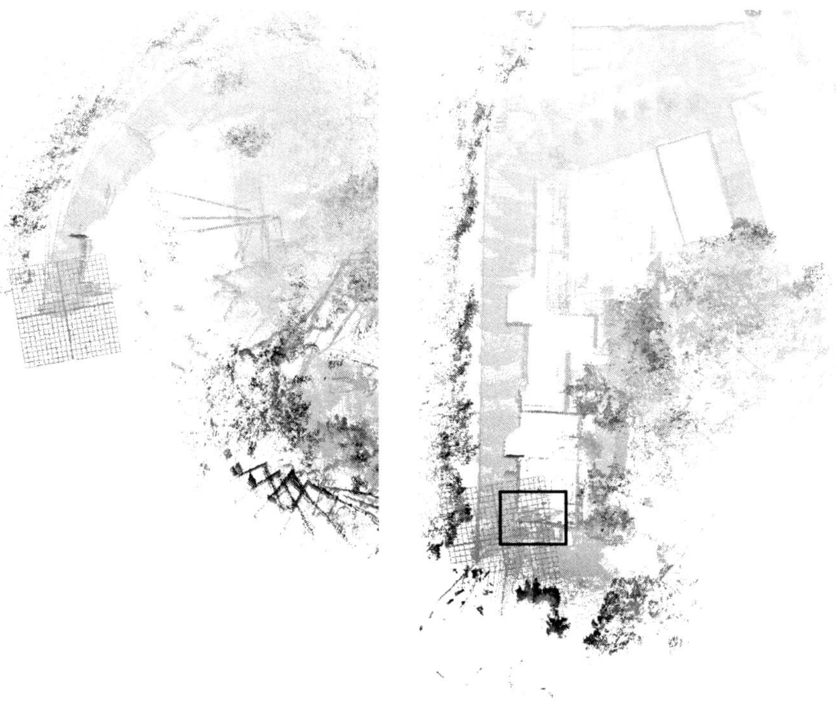

Fig. 6.9 3D model of an experiment to digitize part of the campus of Schloss Birlinghoven campus (top view). Left: Registration based on odometry only. Right: Model based on incremental matching right before closing the loop, containing 62 scans each with approx. 100000 3D points. The grid at the bottom denotes an area of 20×20 m^2 for scale comparison. The 3D scan poses are marked by grey points.

an incorrect minimum. The resulting 3D map would be some mixture of Figure 6.9 (left) and Figure 6.10 (right).

Figure 6.12 shows three views of the final model. These model views correspond to the locations of Kurt3D in Figure 6.2. An updated robot trajectory has been plotted into the scene. Thereby, we assign every 3D scan that part of the trajectory which leads from the previous scan pose to the current one. Since scan matching did align the scans, the trajectory has gaps initially after the alignment (see Figure 6.13).

We calculate the transformation $(\boldsymbol{R}, \boldsymbol{t})$ that maps the last pose of such a trajectory patch to the starting pose of the next patch. This transformation is then used to correct the trajectory patch by distributing the transformation as described in Section 5.1.3. In this way the algorithm computes a continuous trajectory. An animation of the scanned area is available at `http://kos.informatik.uni-osnabrueck.de/download/6Doutdoor/`. The video shows

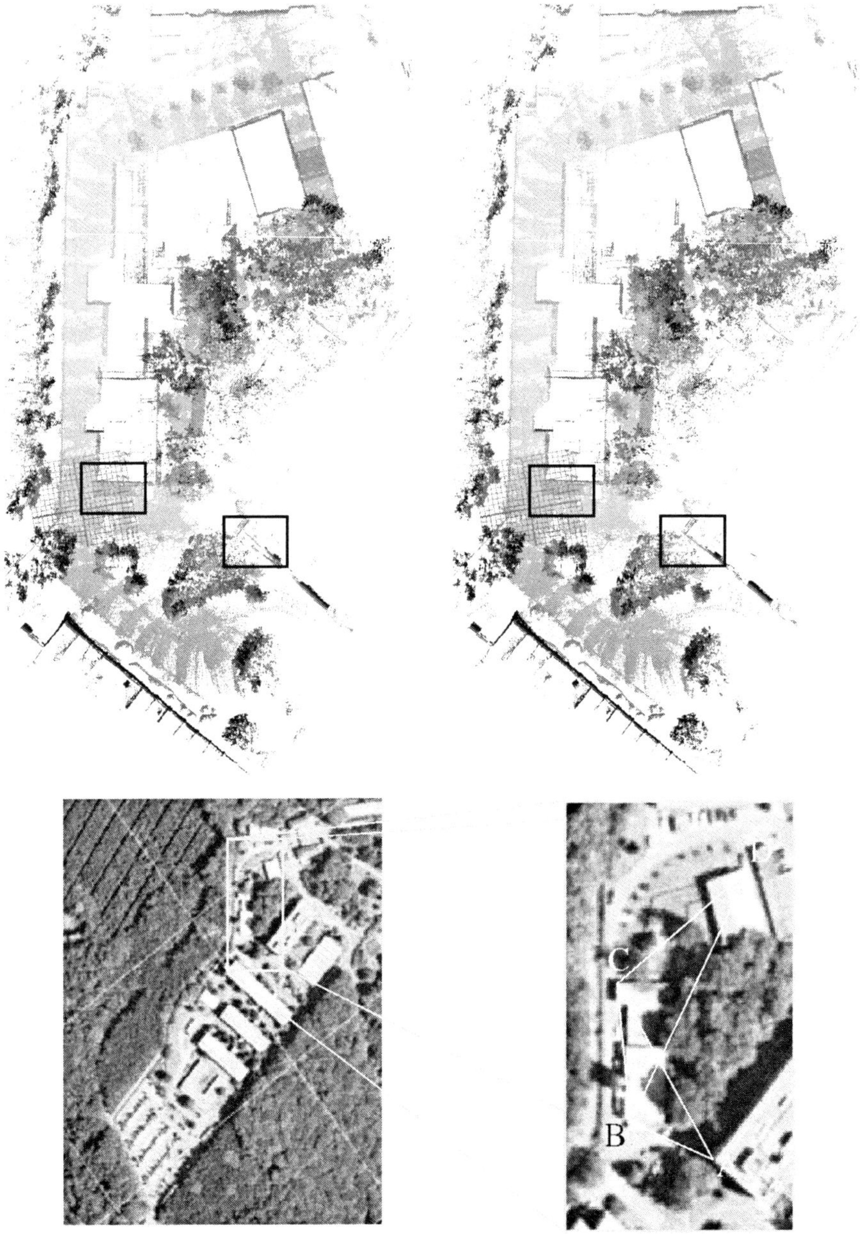

Fig. 6.10 Top left: Model with loop closing, but without global relaxation. Differences to Figure 6.9 right and to the right image are marked. Top right: Final model of 77 scans with loop closing and global relaxation. Bottom: Aerial view of the scene. The points A–D are used as reference points in the comparison in Table 6.1.

Fig. 6.11 Closeup view of the 3D model of Figure 6.10. Left: Model before loop closing. Right: After loop closing, global relaxation and adding further 3D scans. Top: Top view. Bottom: Front view.

the scene along the trajectory as viewed from about 1 m above Kurt3D's actual position.

The 3D scans were acquired within one hour by tele-operating Kurt3D. Scan registration and closed loop detection took only about 10 minutes on a Pentium-IV-2800 MHz, while we did run the global relaxation for 2 hours. However, computing the flight-thru-animation took about 3 hours, rendering 9882 frames with OpenGL on consumer hardware.

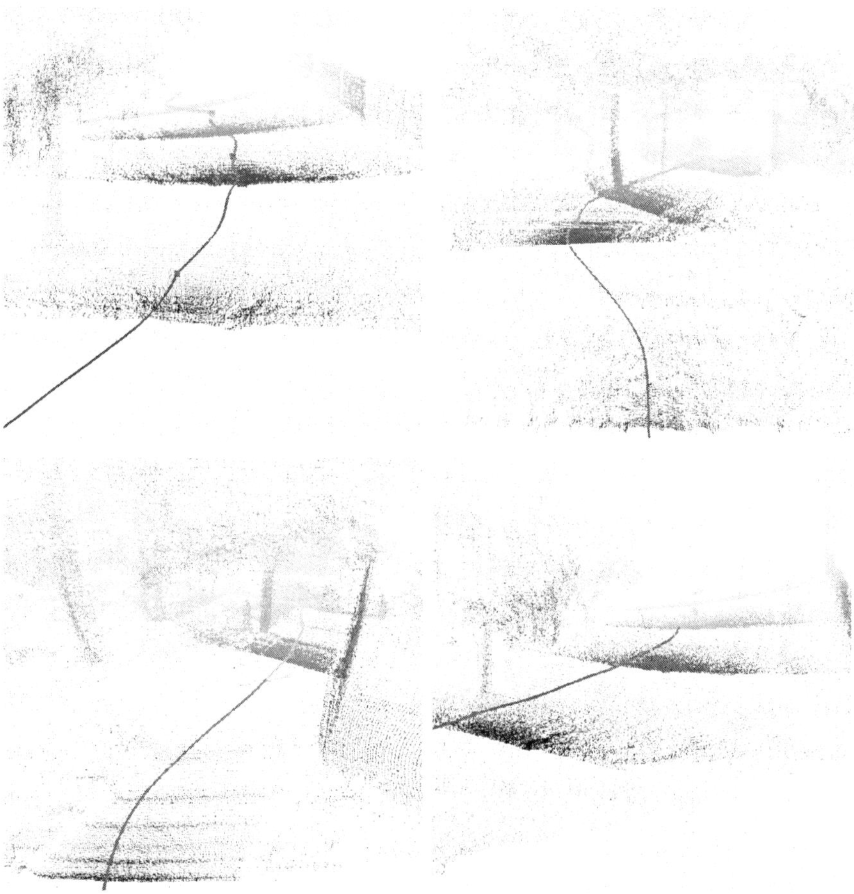

Fig. 6.12 Detailed views of the resulting 3D model, corresponding to the robot locations of Figure 6.2.

Mapping an Eight Shaped Path

The path of the robot led over a bridge (Figure 6.14, right), down the hill, underneath the bridge and back up the hill to the starting point. Figure 6.14 presents the results of matching the 36 3D scans. In addition to sequential scan matching as with ICP, the network contains links connecting the scans taken on the bridge to those taken from below. ICP scan matching shows poor results, i.e., the bridge appears twice in the resulting point cloud. Our algorithm maps the bridge correctly, including the real thickness of the bridge (Figure 6.14). Further experimental results are reported in [13].

Fig. 6.13 The trajectory after mapping shows gaps, since the robot poses are corrected at 3D scan poses.

6.1.6 Kurt3D @ Competitions

RoboCup Rescue

Our 3D mapping algorithms have been tested in various experiments. We participate in RoboCup Rescue competitions on a regular basis. RoboCup is an international joint project to promote AI, robotics and related fields. It is an attempt to foster AI and intelligent robotics research by providing standard problems where a wide range of technologies can be integrated and examined. Besides the well-known RoboCup soccer leagues, the Rescue league is getting increasing attention. Its real-life background is the idea of developing mobile robots that are able to operate in earthquake, fire, explosive and chemical disaster areas, helping human rescue workers to do their jobs. A fundamental task for rescue robots is to find and report injured persons. To this end, they need to explore and map the disaster site and inspect potential victims and suspicious objects. The RoboCup Rescue Contest aims at evaluating rescue robot technology to speed up the development of working rescue and exploration systems [92].

These kinds of competitions allow us to measure the level of system integration and the engineering skills of the teams to be evaluated. It makes high demands on the reliability of the algorithms, since one cannot redo the experiments. A total of 21 robot runs were performed by Kurt3D in the World Championships over the last three years. One major subgoal of such a rescue mission is to create a map of the unstructured environment during the mission time. The test field is a square with 6 meter long sides. Detailed maps of the environment have been presented to the referees right after the robot run. Figure 6.17 shows one such map. With the superimposed grid in Figure 6.17 the referees evaluated the maps facilely.

Fig. 6.14 Top left and right: Closeups on parts of the resulting 3D maps after ICP (left) and LUM (right) scan matching. LUM has shifted the bridge bottom to the correct distance of the bridge surface. Bottom: Photo of the outdoor scene that was mapped.

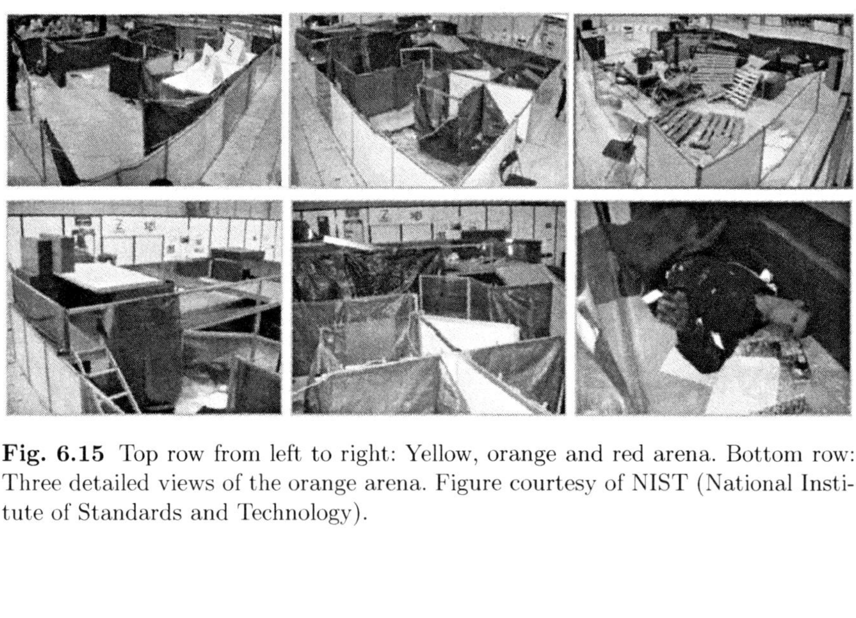

Fig. 6.15 Top row from left to right: Yellow, orange and red arena. Bottom row: Three detailed views of the orange arena. Figure courtesy of NIST (National Institute of Standards and Technology).

Fig. 6.16 Kurt3D at RoboCup Rescue. Top left: 2004 version presented in Lisbon. Top right: 2005 version presented in Osaka features a continously rotating SICK scanner. Bottom: 2006 version accompanied by the robot RTS Crawler.

European Land Robotics Trial

In addition to the RoboCup Rescue competitions, the proposed algorithms have also been tested at the European Land Robotics Trial, ELROB [31]. Please refer to:
`http://kos.informatik.uni-osnabrueck.de/download/Lisbon_RR/`
and
`http://kos.informatik.uni-osnabrueck.de/download/elrob2006/`
for some mapping results.

6.2 Globally Consistent Registration of High Resolution Scans

For this experiment we used a data set from the main square in Horn (Austria), consisting of 13 laser scans. Each scan is composed of 240000 to 300000

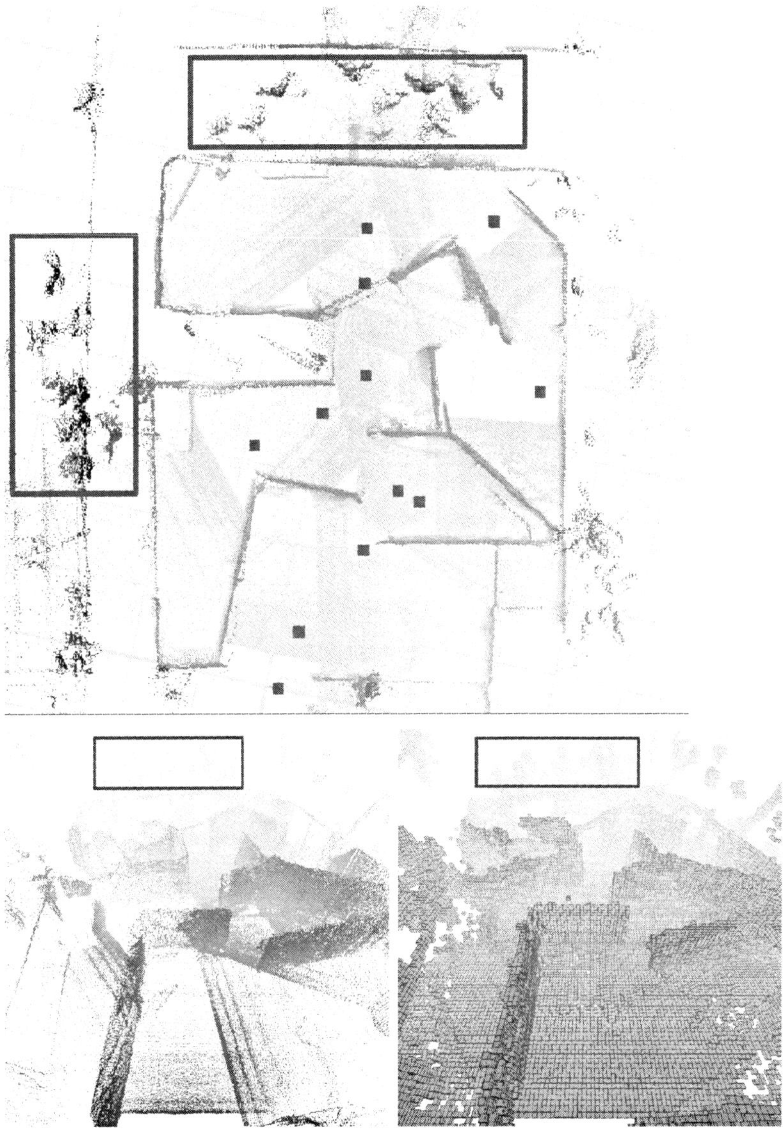

Fig. 6.17 3D maps of the yellow RoboCup arena. The 3D scans include spectators that are marked with a rectangle. Left: Mapped area as 3D point cloud. Middle: Voxel (volume pixel) representation of the 3D map. Right: Mapped area (top view). The points on the ground have been colored in light grey. The 3D scan positions are marked with squares. A $1\,\mathrm{m}^2$ grid is superimposed. Following the ICP scan matching procedure, the first 3D scan defines the coordinate system and the grid is rotated.

Fig. 6.18 Kurt3D presented at ELROB 2006. Similarily to RoboCup the locomotion capabilities play an important role for sucessfully coping with the specific tasks.

points. Scan matching with reduced points took 19 minutes (6 minutes for ICP and 13 minutes for LUM) until convergence, e.g. no scan was moved more than 0.5 cm in one iteration.

Parts of the resulting map can be seen in Figure 6.19 (see `http://kos.informatik.uni-osnabrueck.de/download/Riegl/index.html` for the whole map). From top to bottom, the number of iteration increases, and analogously the precision of the map. Qualitatively, no inconsistencies remain in the bottommost images.

6.3 Mapping Urban Environments – Registration with Dynamic Network Construction

Detecting loops in sets of data helps to build globally consistent maps, because it facilitates distributing larger errors over all scans. Each pose relation gives additional information for improving the calculation of optimal pose estimations. After loop detection, we automatically determine the pose network, using simple distance heuristics. By calculating the pose relations dynamically after each iteration of LUM, we obtain optimized poses, leading

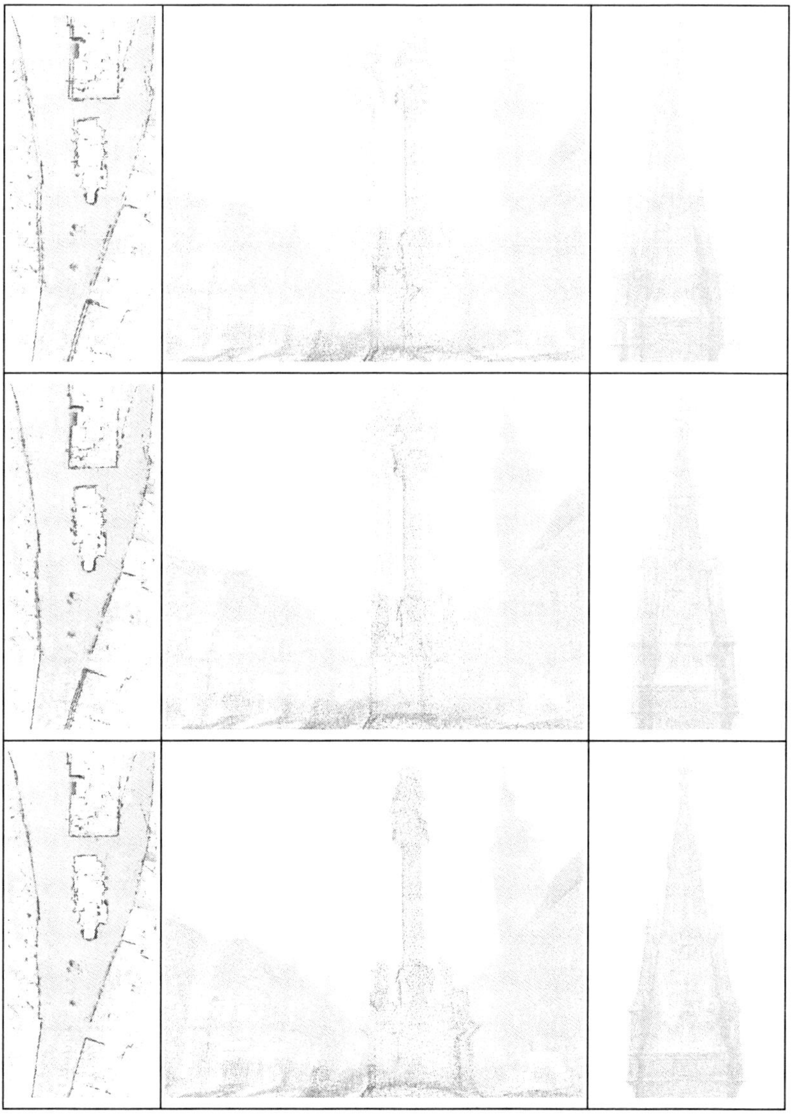

Fig. 6.19 Map improvement during LUM. Main square in Horn (Austria). Data provided by courtesy of RIEGL LMS GmbH [104]. Progress after 1 (top), 30 (middle), 200 (bottom) iterations. Left: Top view, Middle: Monument in the center of the main square, Right: Church spire. Acknowledgements to Nikolaus Studnicka (RIEGL Laser measurement Systems GmbH, Horn, Austria) for providing the data set.

Fig. 6.20 Photos showing the scene presented in Figure 6.19. Data provided by courtesy of RIEGL LMS GmbH [104]. Left: The right part of the middle detailed view. Right: The white-steepled St. Georg church.

iteratively to more accurate networks. Scans that converge towards each other result in new pose relations, while connections between diverging scans disappear.

This is demonstrated on a set of 468 laser scans from a robot run at the Leibniz Universität Hannover while driving a distance of ca. 750 m. Each scan consists of 14000 to 18000 scan points. The convergence limit was set to 0.1 cm movement per pose. Neighbour relations are established between all scans within a distance of less than 7.5 m. But before we describe the mapping result, we introduce the hardware, i.e., the used continuous 3D scanner and the mobile robot Erika. The continuously rotating scanner was also used previously on Kurt3D during the RoboCup Rescue 2005 and 2006.

3D Range Sensor

The sensor that has been employed for the experiments is a fast 3D laser range scanner, developed at the Leibniz Universität Hannover (see Figure 6.21). As there is no commercial 3D laser scanner available that meets the requirements of mobile robots, it is common practice to assemble 3D sensors out of standard 2D laser range sensors and additional servo drives.

The specialties of the RTS/ScanDrive are a number of optimizations that are made to allow fast scanning. One mechanical optimization is the slip ring connection for power and data. This connection allows continuous 360° scanning without the accelerations and high power consumption that are typical for panning systems. Even more important than the mechanical and electrical improvements is the precise synchronization between the 2D laser data, servo drive data and the wheel odometry. Owing to this good synchronization, it is possible to compensate systematic measurement errors and to measure accurate 3D point clouds even with a moving robot. Detailed descriptions of these 3D scanning methods and optimizations are published in [139].

Fig. 6.21 Left: 3D laser range sensor RTS/ScanDriveDuo. It takes a full 3D scan of 32580 points in 1.2 sec. Right: The mobile robot Erika.

Having these optimizations described above, the limiting factor in building faster 3D laser scanner is the maximal number of 13575 (75×181) points that can be measured with a SICK LMS 2xx sensor in one second. The only way of building faster SICK LMS 2xx based 3D scanners is to use multiple 2D measurement devices [128]. For this reason we present the RTS/ScanDriveDuo in this paper. This 3D scanner makes use of two SICK LMS 291 2D laser scanners. Thus the measurement time for 3D scans with 2° horizontal and 1° vertical angle resolution is reduced to 1.2 sec. In this case one 3D scan measured in 1.2 sec consists of 32580 (180×181) 3D points.

The Mobile Robot Erika

The mobile service robot Erika is built from of the Modular Robotics Toolkit (MoRob-Kit). The over all size (L×W×H) of Erika is $95 \times 60 \times 120$ cm. With its differential drive motors, it is able to drive up to 1.6m/s in indoor and urban outdoor environments. The battery capacity is designed to supply the electric wheelchair motors, sensors and a 700 MHz Embedded PC for at least 2 hours or 5 km.

In addition to the 3D laser scanner, the mobile robot is equipped with wheel odometry, a 3 axis gyroscope and a low-cost SiRF III GPS receiver. The measured data of the wheel odometry and the gyroscope are fused to result in the OdometryGyro that is used as the internal sensor for both MCL and SLAM. In contrast to the odometry sensor, the GPS receiver does not influence either of the MCL or the SLAM results. It is only logged to have another laser independent reference.

Fig. 6.22 Sequential ICP registration, from start to first loop detection, in the Hannover data set

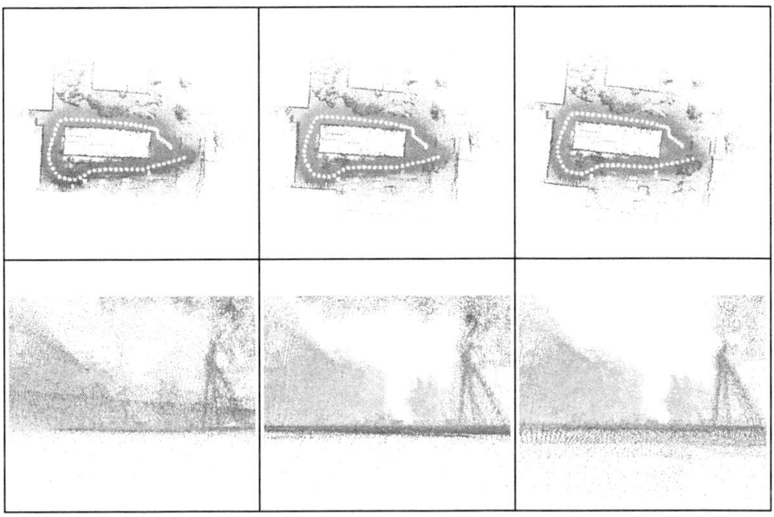

Fig. 6.23 LUM iterations at first closed loop. Close ups made from the same virtual camera position. After few iterations the leaves of the tree are moved behind the camera position (bottom). Left: 1, Middle: 10, Right: 77 iterations.

Mapping Results

Figures 6.22 to 6.26 show the maps after particular steps of scan matching. First, scans are registered iteratively using the ICP algorithm, generating the maps shown in Figure 6.22. Once a loop is detected, LUM is used to achieve global consistency (cf. Figure 6.23). Figure 6.24 shows the map after LUM at the second, third and forth closed loop and Figure 6.26 the final map. While matching the forth loop, the robot path merges through recomputation of the graph, as shown in Figure 6.25.

Ground truth for this data set is not available, therefore no comparison of our final 3D map to a reference 3D model is possible. Wulf et al. developed a method to benchmark our mapping algorithm using Monte Carlo Localization

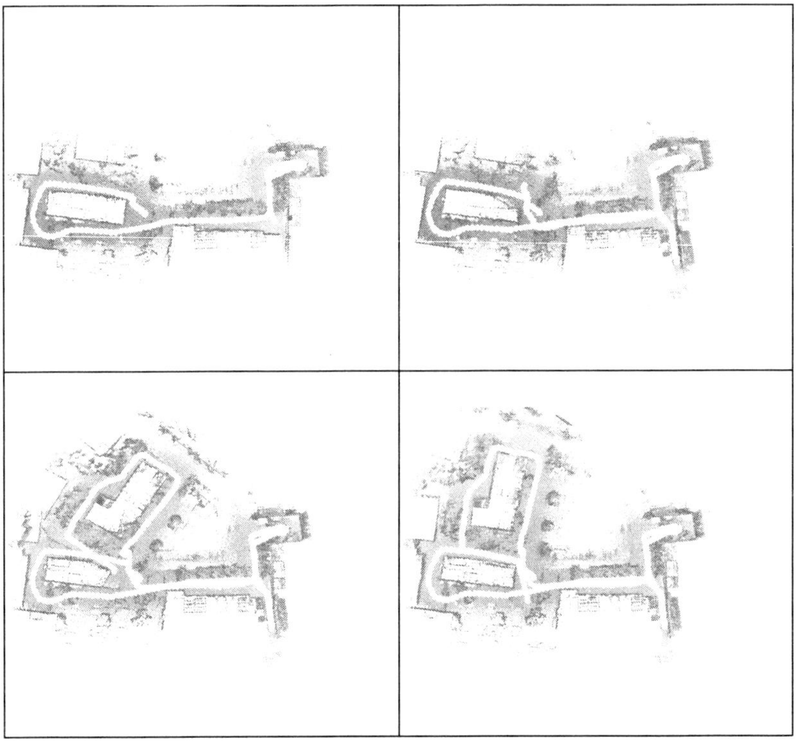

Fig. 6.24 LUM iterations at second, third and fourth closed loop. Top: 2nd and 3rd loop closing. Bottom: 4th loop closing after 1 iteration and after 20 iterations. Notice the merging of the robot path through graph recomputation in the 4th loop closing! (see Figure 6.25 for the corresponding graphs, Figure 6.26 for the resulting map).

in 2D reference maps from the German Land Registry office [138]. Using this novel benchmarking method on a similar data set they demonstrated that our algorithm maps the areas where the loops are closed with higher precision than the remaining parts as described next.

6.4 Benchmarking 6D SLAM

How should results of different SLAM algorithms be compared? An important criterion is clearly the quality of the maps that they would deliver for some test environment. Now what is map quality? Clearly, it is important for a map be consistent, i.e., it should contain no artefacts and doublets but all perceivable structures. Ideally, it should be correct, too, i.e., it should represent metrical information about the environment in a correct way.

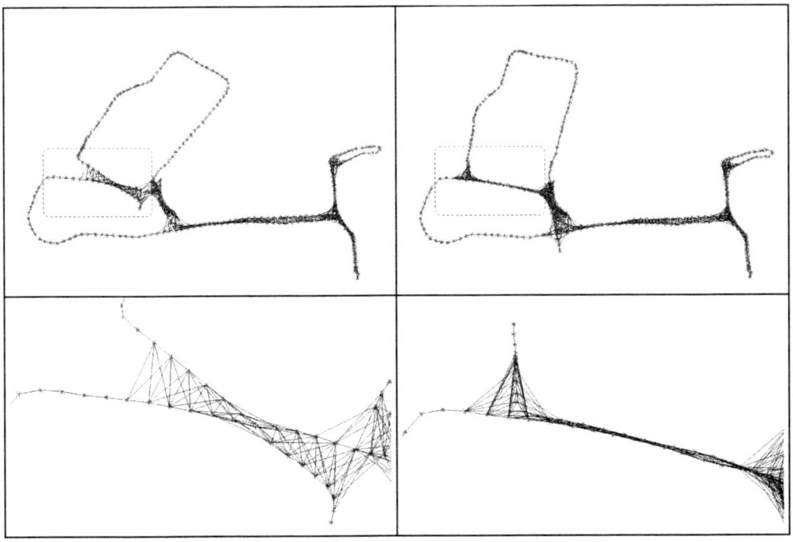

Fig. 6.25 Top: Graphs corresponding to the two alignments of Figure 6.24 (bottom). Bottom: Zoom into the boxed areas.

While it is cumbersome but feasible to determine correctness of typical planar indoor maps (use a yardstick for making independent measurements), it is practically impossible to get this type of independent ground truth information for large 3D outdoor scenes. So on the one hand, generating 3D maps of large environments, using 3D or 6D poses, has received much attention recently, e.g., [19, 128, 97] Yet, on the other hand, a framework for benchmarking these large experiments is still missing. Well known SLAM sites offer only data sets, e.g., *Radish: The Robotics Data Set Repository* [63] or algorithms, e.g., OpenSLAM [113] and no ground truth data. So, given a map generated by some SLAM algorithm, we may judge whether the map "looks good", i.e., is not obviously inconsistent; but there is no way to measure its quality with respect to independent reference data, let alone to determine its consistency with respect to ground truth. So comparing systematically the maps resulting from different SLAM approaches is a futile attempt. Mapping environments by conventional means, i.e., using markers and geodetic surveying equipment, ends up with maps containing similar measures of uncertainty, and such guarantees are required by many mapping customers, e.g., applications in facility management and architecture, or in construction and maintenance of tunnels, mines, and pits. In contrast, current SLAM technology cannot come up with performance measures.

This section presents a method for generating independent reference data for benchmarking outdoor SLAM approaches. In particular, we are aiming for full 6D SLAM approaches. Our procedure makes use of an independently available, accurate environment map (provided by the land registry office), a

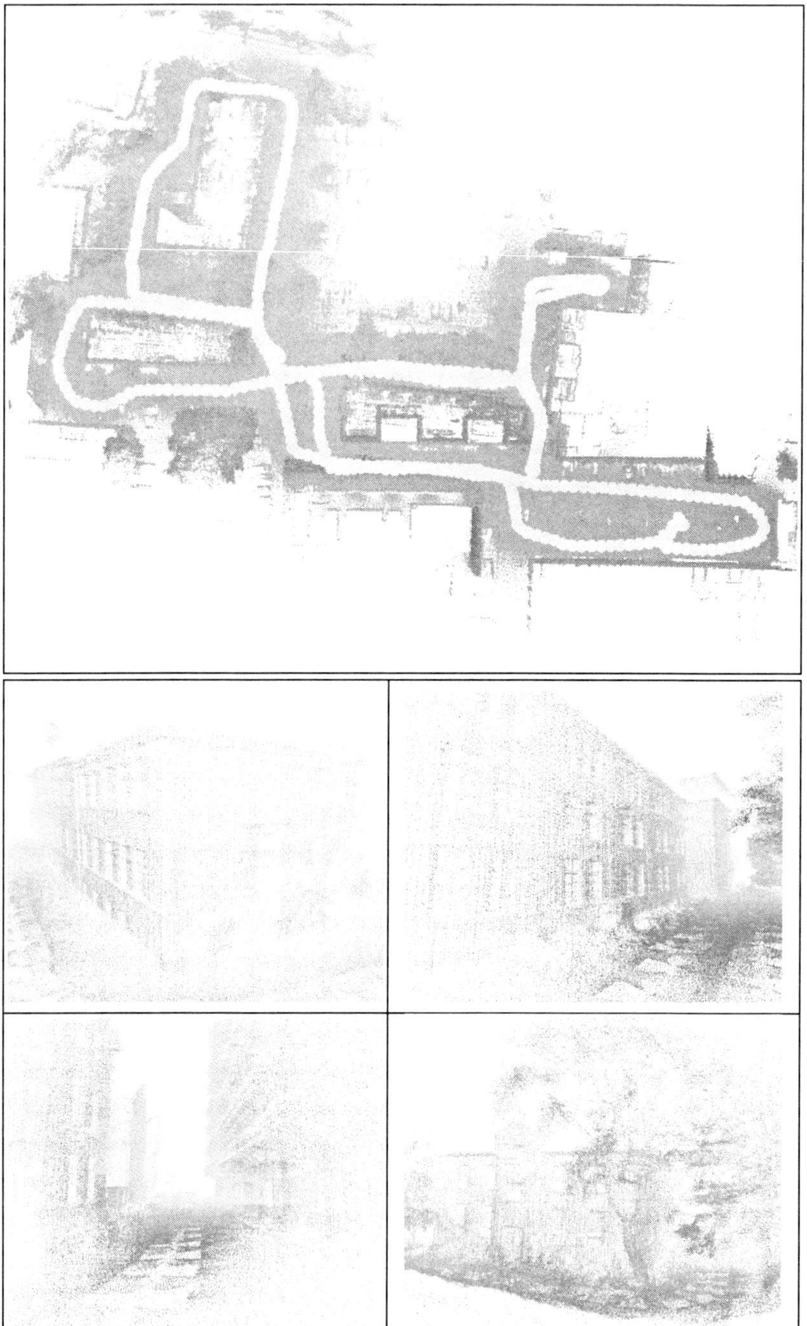

Fig. 6.26 Final map. Top: Top view including the final trajectory. Bottom: Four details rendered from the Hannover scene.

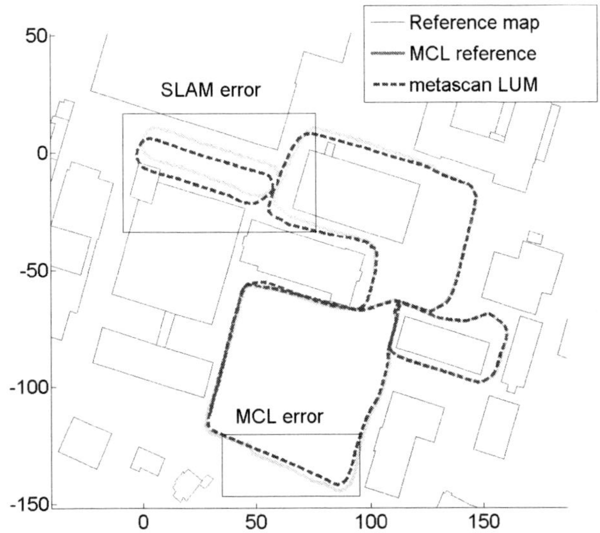

Fig. 6.27 Performance measurement of SLAM algorithms using reference maps and MCL in an urban outdoor environment (University of Hannover campus, 3D scan indices 1-700). Distances are given in meters.

Monte Carlo Localization (MCL) technique that matches sensor data against the reference map, and manual quality control.

To demonstrate the use of the benchmarking method, we apply it for determining the quality of a family of 6D SLAM variants, derived from Figure 5.1 (cf. page 82). The data used for the respective experiments has been recorded using a mobile robot that was steered manually in natural terrain, over asphalt roads, sidewalks, cobblestone etc., with a fast 3D scanner gaging the environment. A sketch of the demo environment is given in Figure 6.27.

6.4.1 Ground Truth Experiments

Doing experiments with ground truth reference, researchers aim at measuring the objective performance of a dedicated algorithm. Based on this benchmark it is possible to give an experimental proof of the effectiveness of a new algorithm. Furthermore, measuring the performance of an algorithm allows it to be optimized and compared to other existing solutions.

Benchmarking is a common instrument in science. Good examples for successful performance measurements in computer science are available in the computer vision community. There are several projects that aim at providing image databases for other researchers [60] [126]. These image databases are supplemented by ground truth images and algorithms that calculate performance metrics. In doing so, the community is able to make progress and to document its progress in fields like image segmentation and object recognition.

Unfortunately, this kind of performance measurement is not widely spread in the robotics community. Even though there are several ways of comparing the performance of robotic algorithms and systems, one basic step is to provide experimental data and results to other research groups. Up to now this is only done by small projects [63, 113] or by individual researchers. Another way of comparing robotic systems is competitions like RoboCup [29], ELROB [31] or the Grand Challenge [23]. This kind of competitions allows the level of system integration and the engineering skills of a certain team to be ranked, but it is not possible to measure the performance of a subsystem or a single algorithm.

Objective benchmarking of localization and mapping algorithms is only achieved by comparing experimental results against reference data. The practical problem is to generate this ground truth data. In computer vision, ground truth data is either available for synthetic images, or needs to be manually labelled. In case of mobile robot navigation one way of gathering ground truth data is to use precise global positioning systems (RTK-GPS) [52]. Unfortunately, this data is only available in open outdoor environments and not for urban outdoor environments or indoor environments. Another possibility is to use complex external measurement setups.

Simulation yields another benchmarking method for robotic algorithms, enabling researchers to perform experiments in defined conditions and to repeat these experiments. However, real life differs from simulation. Experiments involving sophisticated sensors such as cameras or laser scanners can only be simulated up to a certain level of accuracy, e.g., capturing environments must regard surface properties such as material, local structures and reflections. Therefore, using real robotic data sets is favoured for benchmarking.

6.4.2 The Benchmarking Technique

This book introduces a new benchmarking technique for SLAM algorithms. The benchmark is based on the final SLAM results and a reference position that is obtained independently of the SLAM algorithm under test.

As highly accurate RTK-GPS receivers can not be used in urban outdoor environments, we present a technique that is based on surveyed maps as they can be obtained from the German land registry offices. The process of generating these ground truth reference positions can be divided into a Monte Carlo Localization (MCL) step that matches the sensor data to the highly accurate map, and a manual quality control step to validate the MCL results.

As the SLAM algorithm under test and the MCL algorithm use the same sensor data, the SLAM results and the reference positions are not completely independent. On the other hand, global localization algorithms and incremental localization and mapping algorithms work differently. Incremental mapping algorithms like odometry and SLAM can suffer from accumulating errors and drift effects. However, pure localization algorithms eliminate these errors by continuously matching to a given accurate map. Therefore, the remaining error of the manually supervised reference position is at least an order of magnitude smaller then the discussed SLAM errors.

Reference Map

As part of their geo information system (GIS), the German land registration offices feature surveyed data of all buildings within Germany. The information about these building is stored in vector format in the so-called "*Automatisierte Liegenschaftskarte*" (ALK). The vector format contains lines that represent the outer walls of solid buildings. Each line is represented by two points with northing and easting coordinates in a Gauss-Krueger coordinate system. The upper error bound of all points stored in the ALK is specified to be 4 cm. Up to now, no further details about doors, windows or balconies are available.

If no reference map is at hand, one can produce one from calibrated areal photographs or satellite images. The accuracy of such maps is usually in the range of 10 cm.

Monte Carlo Localization

The Monte Carlo Localization (MCL) is a commonly used localization algorithm that is based on particle filtering [35]. As the theory of MCL is well understood we focus on the sensor model that is used to match the 3D sensor data to the 2D reference map.

The key problem of matching a 3D laser scan to a 2D map is solved by using a method called *Virtual 2D Scans* [136]. The method splits up into two steps. The first step reduces the number of points in the 3D point cloud. The reduction step is based on the assumption that the reference map presents plain vertical walls. For this reason all 3D measurement points that do not belong to plain vertical surfaces need to be removed (Figure 6.28). [137] describes a sequence of 3D segmentation and classification algorithms that is used to do this reduction in urban outdoor environments. By this means, the ground floor, vegetation, and small objects are removed from the 3D data. Measurement points on the outer walls of buildings and on other unmapped vertical obstacles remain. Thus, cars, people etc, are removed in urban scenes, since the scanner usually acquires data points above these objects and the *Virtual 2D Scan* contains data points on buildings.

Having this reduced 3D point cloud, the second step of the *Virtual 2D Scan* method is a parallel projection of the remaining 3D points onto the horizontal plane. After this projection the y coordinate contains no information and can be removed. By this means, the *Virtual 2D Scan* has got the same data format

Fig. 6.28 Classified 3D point cloud of an urban scene, featuring a building and bicycles in front. (ground pounds -light gray; plain vertical surface points - dark gray; others - black)

as a regular 2D scan. Thus it can be used as input data of a regular 2D MCL algorithm. To reduce the computational complexity of the successive MCL algorithm the remaining measurement points are randomly down sampled. Experimental results show that less than 100 measurement points are needed for sufficient localization. Thus the average 3D point cloud with about 30,000 measurement points is reduced to a *Virtual 2D Scan* with only 100 points without losing information that is needed for localization in urban outdoor environments.

Due to the 2D nature of the reference map and the used 2D MCL algorithm, only the 3 DoF pose $\boldsymbol{p}^{\mathrm{REF}} = (x, y, \theta_z)$ of the robot can be estimated. There is no reference information on the robot's elevation y. Furthermore, the roll and pitch components θ_z and θ_x of the 6 DoF robot pose can not be estimated with this 2D method. These angles need to be measured and compensated with a gyro unit before the generation of the *Virtual 2D Scans.*

Manual Quality Control

As it is described above the MCL is used to generate the reference path from the given map and sensor readings. But the output of the MCL can not be stated as ground truth in general. Like all measured signals the reference path has got an error. The absolute error depends on the error of the reference map, the sensor errors and the convergence and precision of the MCL algorithm. To be able to use the reference path as ground truth in our benchmark, the absolute error of the reference path needs to be at least one order of magnitude smaller than the error of the SLAM algorithms under test.

To ensure the quality of the reference path a manual quality control step is integrated into the benchmark procedure. In this step a human supervisor monitors the sensor data, the particle distribution and the matching results to decide if the calculated reference path can be used as ground truth. The manual quality control decides which parts of the path are used for benchmarking, but the manual supervisor is not taking any influence on the sensor data of the MCL algorithm.

First, the sensor data needs to be checked for a sufficient number of landmarks, namely, walls as given in the reference map. If there are not enough landmarks in the environment and thus in the *Virtual 2D Scans*, the MCL results depend only on odometry and are therefore inaccurate.

Second, the numerical condition of the particle filter needs to be monitored. As a particle filter only presents a sampled belief, an efficient distribution of the finite number of particles is essential for correct operation and precise results.

If these two criteria are fulfilled the remaining error is white noise with zero mean and a small standard deviation depending on the map, the 3D sensor and the discretization of the particle filter.

Benchmark Criteria

Up to this point, the MCL positions and SLAM positions are given in different coordinate systems. The MCL positions are given in the global Gauss-Krueger coordinate system of the reference map and the SLAM positions are given in a local coordinate system that is centered at the robot's start position. To be able to compare the positioning results, the SLAM positions need to be transformed into the global coordinate system based on the known start position.

Having the trusted MCL reference $\boldsymbol{p}^{\mathrm{REF}}$ and the SLAM results $\boldsymbol{v}^{\mathrm{SLAM}}$ in the same coordinate system, it is possible to calculate objective performance metrics based on position differences. The first metric based on the 2D Euclidean distance between the SLAM and MCL position

$$e_i = \sqrt{(x_i^{SLAM} - x_i^{REF})^2 + (z_i^{SLAM} - z_i^{REF})^2}. \tag{6.1}$$

The second metric is based on the difference between the SLAM and MCL orientation

$$e_{\theta,i} = |\theta_{y,i}^{SLAM} - \theta_{y,i}^{REF}|. \tag{6.2}$$

As the MCL position is given in 3 DoF only, the robot's elevation, roll and pitch angles cannot be tested.

To compare the performance of different SLAM algorithms on the same data set, it is possible to calculate scores like the standard deviation

$$\sigma = \sqrt{\frac{1}{n+1}\sum_{i=0}^{n} e_i^2},$$

and the error maximum

$$e_{max} = \max e_i.$$

Of course, these statistic tests can be done analogously on the orientation errors $e_{\theta,i}$, resulting in the scores (σ_θ and $e_{\theta,max}$).

6.4.3 Experimental Results

Experimental Setup

The presented experiment was carried out at the campus of the Leibniz Universität Hannover. The experimental robot platform that was used to collect the data was manually driven on the 1.242 km path closing a total of 5 small and large loops. On this path 924 full 3D scans were collected at an average robot speed of 4 km/h and a maximum speed of 6 km/h. In addition to the 3D laser data, wheel odometry and fused wheel/gyro odometry were stored with a data rate of 10 Hz. And the position fixes of a low-cost GPS were logged with 1 Hz.

The Reference Path

The section of the ALK that is used as the reference map contains 28 buildings represented by 413 line segments. To avoid huge coordinate numbers a constant offset of 5806400 m northing and 3548500 m easting is subtracted from all Gauss-Krueger coordinates. This offset corresponds to the position 52°23′58″ north, 9°42′41″ east in WGS84 coordinates.

The reference path is calculated with a Monte Carlo Localization based on 3D laser data and wheel/gyro odometry. The particle filter with 200 samples runs in parallel to the data acquisition and logging threads on a Pentium III 700 MHz processor. The localization results are plotted as a solid gray line in Figure 6.27.

For the manual quality control the MCL results can be displayed offline. As a result of the manual quality control it came out that the reference

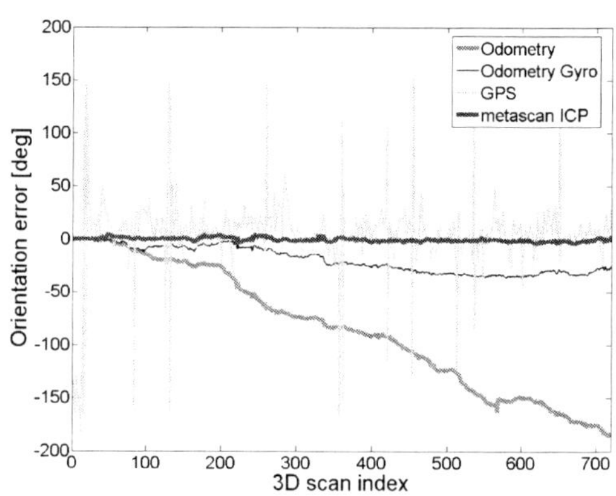

Fig. 6.29 Orientation errors. Comparing internal sensors measurements, GPS headings and *metascan ICP* matching with orientations computed by MCL localization. The x-axis represents the 3D scan index, roughly corresponding to the position at the robot path. The orientation errors have been computed using Eq. (6.2).

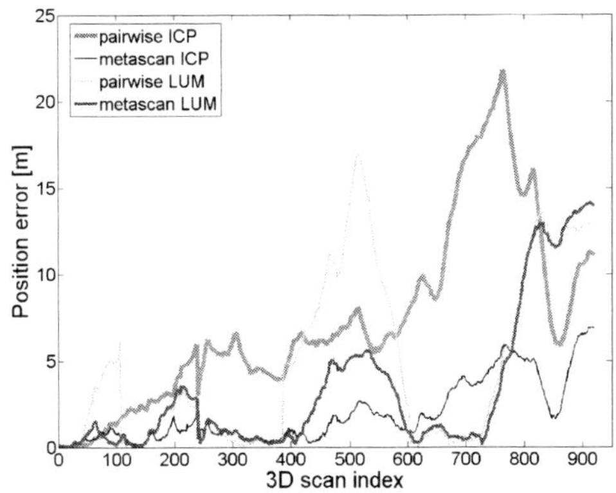

Fig. 6.30 Position errors. Comparison of different mapping strategies. The position errors have been computed using Eq. (6.1).

path for 3D scan index 198 - 241 can not be used for benchmarking as there are not enough landmarks visible in the 3D laser scan. As it can be seen in Figure 6.27 (MCL error box) the MCL results drift away from the true position. From index 236 on the 3D scan and thus the *Virtual 2D Scan* contains new landmarks. As of index 242 the MCL is converged to the true position again. Thus, the manual quality control decides that the reference path from index 1 to 197 and 242 to 924 can be used as ground truth in the benchmarking process. The remaining error of the reference path is estimated to be white noise with zero mean and a standard deviation of 10cm.

Mapping Results

Mapping with Internal Sensors and GPS. Since any sensor is inaccurate, the maps generated using internal sensors for pose estimation are of limited quality, as has been demonstrated many times before. For odometry and the gyro based localization, the error for orientation and position are potentially unbounded. However, since paths usually contain left and right turns, these errors partially balance. The GPS shows problems close to buildings, where the orientation is poorly estimated and the position error reaches its maximal value. Figure 6.29 shows the orientation errors of the internal sensors in comparison to ICP scan matching.

Mapping with ICP. Mapping with ICP was done using two different methods, namely *pairwise ICP* and *metascan ICP*. The latter method outperforms *pairwise ICP* since it considers all previously acquired 3D scans leading to slower error accumulation. Figure 6.30 shows the scan matching errors in comparison to methods using explicit loop closure that are described next.

Mapping with ICP and Global Relaxation. The performance of the methods *pairwise LUM, metascan LUM* have also been evaluated. As expected, loop closing reduces the position error at the positions, where the

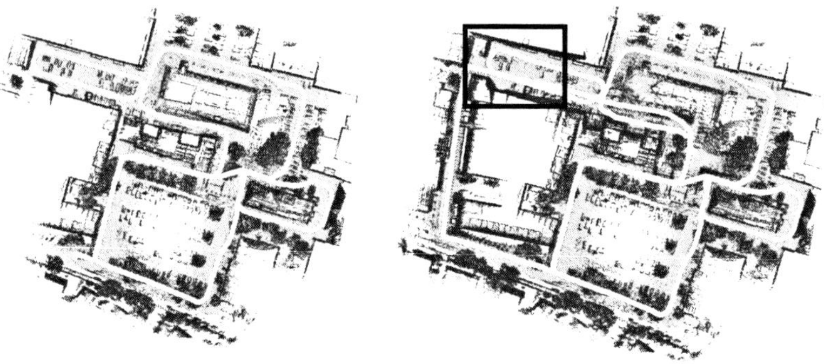

Fig. 6.31 Final 3D map using the metascan LUM strategy. 3D points are classified as ground (light gray) and object points (dark gray). The trajectory is denoted in white. Left: Registration of the first 720 3D laser scans into a common coordinate system. Global relaxation leads to a consistent map. Right: The accumulated elevation errors on the remaining path (3D Scan 700 to end) prevents loop closing (black rectangle). Due to that parts of the map are inconsistent. A detailed view of the black rectangle is provided in Fig 6.34.

loop is closed to approximately zero, e.g., Figure 6.30 at scan index 100, where the first loop was closed and at the indices 300–400 and 600–700. At these locations, the Lu/Milios style SLAM methods outperform the *pairwise ICP* and *metascan ICP* methods and the absolute error is close to zero. However, since global relaxation produces consistent maps the error in between the loop closured might be larger than the one obtained without global relaxation. Having consistent maps does not necessarily imply correct maps. *pairwise LUM*, and *metascan LUM* may also fail, if the loop cannot be closed. This case occurs in our experiment in the final part of the trajectory, i.e., when the scan index is greater than 700 (cf. Figure 6.30 and Figure 6.34). This last loop was not detected by the threshold method described in section 5.2.2.

Table 6.2 Position errors [m]

method	σ	e_{max}
Odometry	55.1	261.2
OdometryGyro	64.7	250.1
GPS	5.8	95.1
pairwise ICP	5.2	21.8
metascan ICP	1.6	6.6
pairwise LUM	4.9	17.0
metascan LUM	3.8	13.8

Fig. 6.32 3D view corresponding to Figure 6.31, left. Two close-up views are provided and the corresponding camera positions are shown. The trajectories are given in Figure 6.33.

Finally, Table 6.2 and 6.3 compare all localization/mapping methods. Figure 6.31 shows the final map generated with *metascan LUM*. The left part contains the first 720 3D scans that have been matched correctly, whereas the right part contains all scans including the errors, due to the undetected loop. Figure 6.32 shows a 3D view of the scene including two close-up views.

6.4.4 Justification of the Results

To validate our experimental methodology, i.e., to generate ground truth reference positions using MCL as described, we match the acquired 3D scans with a 3D map generated from the 2D reference map. For this, the 2D map

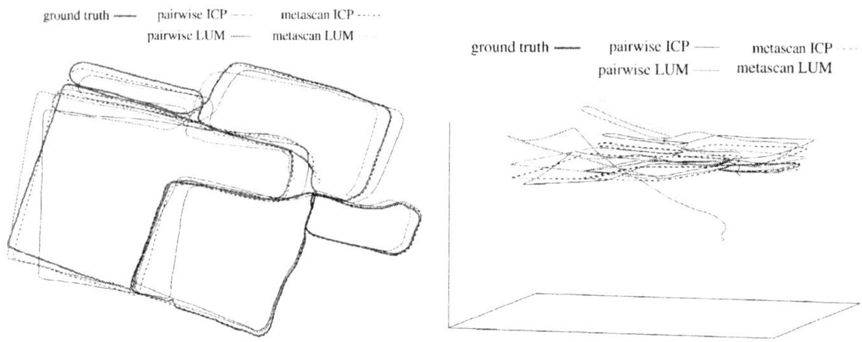

Fig. 6.33 Computed trajectories in comparison. Right: Top view. Left: 3D view. Due to the accumulated height error the last loop closing is not possible and the computed error score of the SLAM algorithms with global relaxation is larger than *metascan ICP* (cf. Table 6.2).

Table 6.3 Orientation errors [deg]

method	σ_θ	$e_{\theta,max}$
Odometry	77.2	256.6
OdometryGyro	15.1	56.7
GPS	27.3	171.0
pairwise ICP	6.3	17.7
metascan ICP	2.4	11.8
pairwise LUM	5.2	22.8
metascan LUM	4.3	21.2

is extrapolated as a 3D map, cuboids representing the boundaries of the buildings. Figure 6.35 shows the final map with the point clouds representing the buildings.

This 3D map, which is a 2D map extended by some height, is used for comparison using the following three strategies:

1. ICP is used for matching every single 3D scan with the point cloud based on the 2D map. As it turns out, this method can only be applied to the first 200 scans, since the map does not cover the whole robot path. In comparison, MCL successfully deals with this problem by applying the motion model to the particles until scan matching is possible again. This method is referred to as *map ICP (first part)*.
2. ICP is used for matching a 3D scan with the metascan consisting of the 3D points from the map and all previous acquired and registered 3D scans. The method will be called *metascan map ICP*.
3. The previous method is used with the extension that points classified as ground are not included, i.e., only the dark grey points are used for

Fig. 6.34 3D view of the problematic loop closure in Figure 6.31 (right, black rectangle). Loop closing is not possible due to accumulated elevation errors.

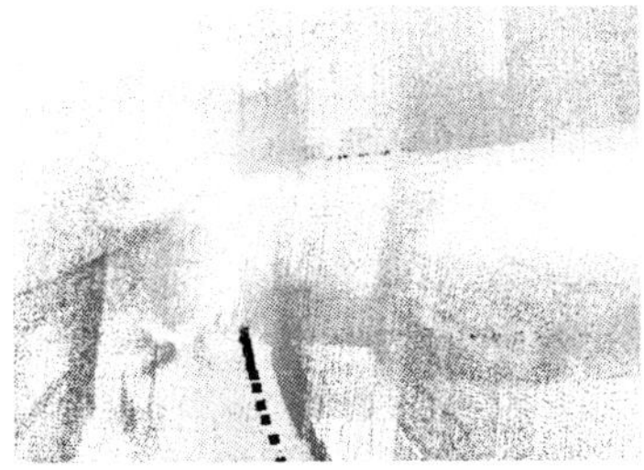

Fig. 6.35 Mapped 3D scans overlaid with 3D cuboids based on the 2D reference map

computing point correspondences and the transformation. Thus it is called *metascan map ICP w/o ground*. It is expected that this restriction results in better ICP convergence.

Figures 6.36 and 6.37 compare the additional localization/mapping methods. Table 6.4 and 6.5 give quantitative results. It turns out that these justification methods behave similar to MCL and produce comparable results, i.e., the MCL trajectory differs only by statistical noise from the trajectories produced by ICP scan matching using a 3D point cloud derived from the map.

The evaluation above regards only three DoF, since only the x- z and θ_y parts of the pose are benchmarked. For evaluating all six DoF one needs a

Table 6.4 Position errors [m]

method	σ	e_{max}
map ICP (first part)	0.3	1.77
metascan map ICP	0.38	1.79
matascan map ICP wo ground	0.38	2.98

Table 6.5 Orientation errors [deg]

method	σ_θ	$e_{\theta,max}$
map ICP (first part)	1.24	4.00
metascan map ICP	1.40	6.37
matascan map ICP wo ground	1.38	6.32

3D referecne map, acquired by an independent measurement. Such a measurement is given by 3D laser scans acquired from a flying airplane. Figure 6.39 shows such a laser scan. Note the white areas are scanned roofs of buildings, that have been clipped. Since vertical structures, like walls have not been scanned, the aerial laser scan has been extended by the 2D reference map, extrapolated to the third dimension. Figure 6.38 left shows such a 3D map. The right part shows the 3D map superimposed by the 3D laser scans acquired by the robot Erika.

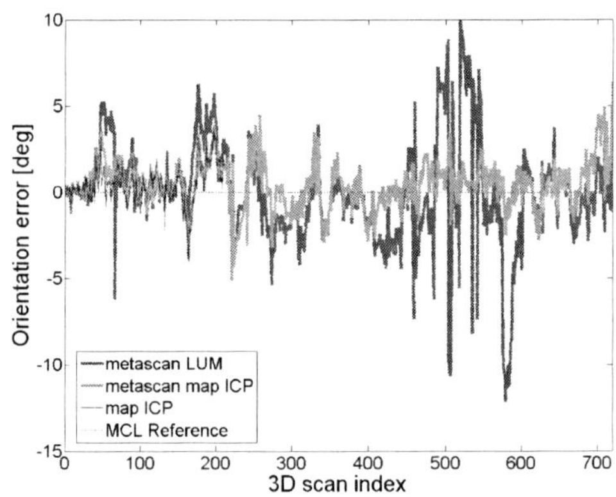

Fig. 6.36 Orientation errors. Comparing metascan LUM, metascan map ICP and map ICP (first part) with orientations computed by MCL localization. The x-axis represents the 3D scan index, roughly corresponding to the position at the robot path.

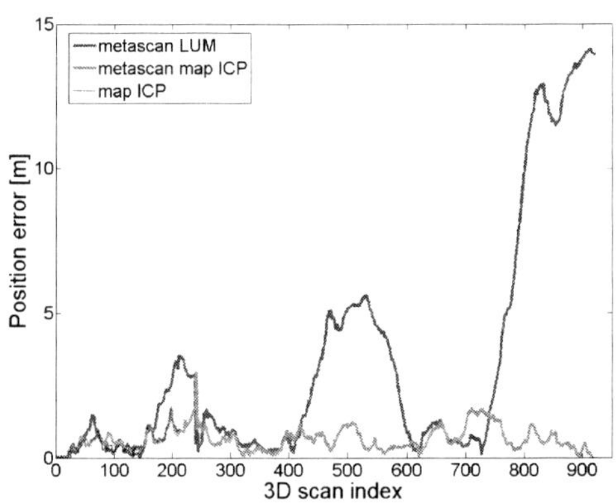

Fig. 6.37 Position errors (Euclidean distance to MCL localization). Comparing metascan LUM, metascan map ICP and map ICP (first part).

Fig. 6.38 Detailed views of Figure 6.39. Left: Aerial 3D scan extended by the extrapolated 2D reference map. Right: Aereal 3D scan superimposed to the point cloud acquired by the robot.

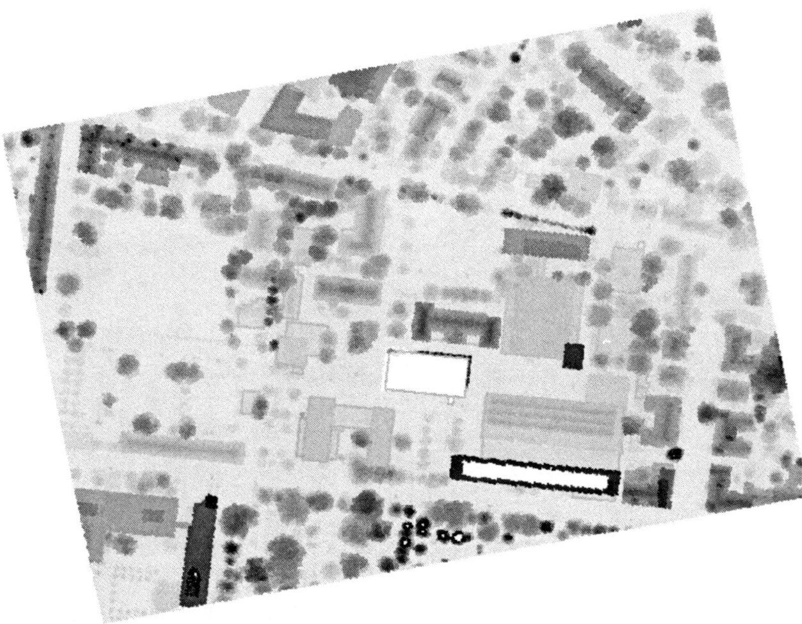

Fig. 6.39 3D laser scan acquired from an airplane. Data set courtesy of Claus Brenner, Leibniz University Hannover.

6.5 Further Applications

6.5.1 Mapping Abandoned Mines

This section describes an algorithm for acquiring volumetric maps of underground mines. Mapping underground mines is of enormous societal importance [84], as a lack of accurate maps of inactive, underground mines poses a serious threat to public safety. According to a recent article [6], "tens of thousands, perhaps even hundreds of thousands, of abandoned mines exist today in the United States and worldwide. Not even the U.S. Bureau of Mines knows the exact number, because federal recording of mining claims was not required until 1976." The lack of accurate mine maps frequently causes accidents, such as a recent near-fatal accident in Quecreek, PA [95]. Even when accurate maps exist, they provide information only in 2D, which is usually insufficient to assess the structural soundness of abandoned mines. Accurate 3D models of such abandoned mines would be of great relevance to a number of problems that directly affect the people who live or work near them. One is subsidence: structural shifts can cause collapse on the surface above. Ground water contamination is another problem of great importance, and knowing the location, volume, and condition of an abandoned mine can be highly informative in planning and performing interventions. Accurate volumetric maps are also of great commercial interest. Knowing the volume of the material already removed from a mine is of critical interest when assessing the profitability of re-mining a previously mined mine.

Hazardous operating conditions and difficult access routes suggest that robotic mapping of abandoned mines may be a viable option to traditional

Fig. 6.40 Superposition of the aerial laser scan (black) with the robotc 3D scans (grey)

manual mapping techniques. The idea of mapping mines with robots is not new. Past research has predominantly focused on acquiring maps for autonomous robot navigation in active mines. For example, Corke and colleagues [20] have built vehicles that acquire and utilize accurate 2D maps of mines. Similarly, Baily [5] reports 2D mapping results of an underground area using advanced mapping techniques.

CMU Mine Mapping

The algorithms have been tested with 3D scans taken in the Mathies mine, Pittsburgh, PA. The experiments reported in this section utilize data by CMU's mine mapping robot Groundhog [30,123], which relies on 2D mapping for explorating and mapping of abandoned mines. While Groundhog and a related bore-hole deployable device "Ferret" [84] utilizes local 3D scans for navigation and terrain assessment, none of these techniques integrates multiple 3D scans and generates full volumetric maps of abandoned mines.

The Groundhog Robot. The robot has been deployed and built by the CMU Mine Mapping Team [30, 123]. Groundhog's chassis unites the front halves of two all terrain vehicles, allowing all four of Groundhog's wheels to be both driven and steered. The two Ackerman steering columns are linked in opposition, reducing Groundhog's outside turning radius to approximately 2.44m. A hydraulic cylinder drives the steering linkage, with potentiometer feedback providing closed-loop control of wheel angle. Two hydraulic motors coupled into the front and rear stock ATV differentials via 3:1 chain drives result in a constant 0.145 m/sec velocity. When in motion, Ground hog consumes upwards of 1kW, where processing and sensing only draw 25W and 75W respectively. Therefore, time spent sensing

Fig. 6.41 The Groundhog robot

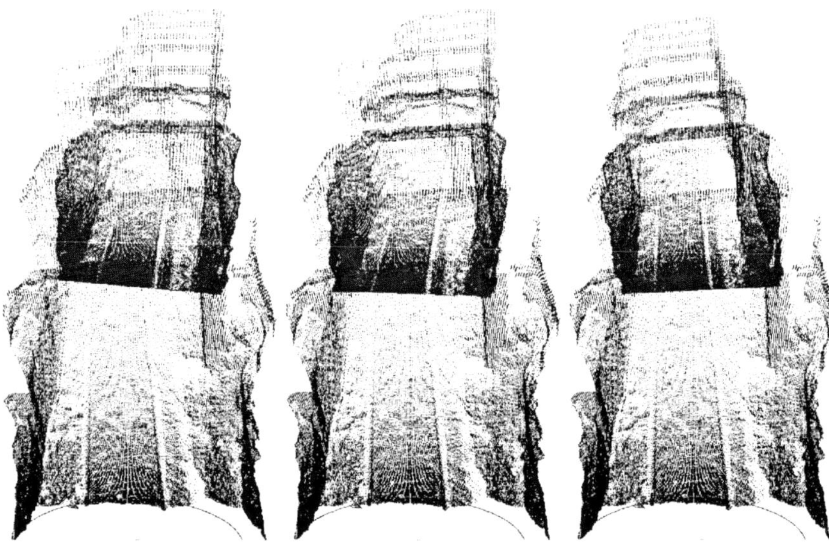

Fig. 6.42 Left: Initial odometry-based pose of two 3D scans. Middle: Pose after five ICP iterations. Right: final alignment.

and processing has minimal impact on the operational range of the robot. The high power throughput combined with the low speed of the robot means that Groundhog has the torque necessary to overcome the railway tracks, fallen timbers, and other rubble commonly found in abandoned mines. Equipped with six deep-cycle lead-acid batteries, and in later experiments with eight such batteries, Groundhog has a locomotive range greater than 3km.

For acquiring 3D scans in a stop and go fashion, Groundhog is equipped with tiltable SICK laser range finders on either end. The area of $180°(h) \times 60°(v)$ is scanned with the horizontal resolution of 361 pts. and vertical of 341 pts. A slice with 361 data points is scanned in 26ms by the 2D laser range finder (rotating mirror device). Thus a scan with 361×341 data points needs 8.9 seconds. Figure 6.42 shows an example scan of the mine.

The Mathies Mine Experiment. Groundhog's development began in the Fall of 2002 by the CMU Mine Mapping Team [30,123]. The robot was extensively tested in a well-maintained inactive coal mine accessible to people: the Bruceton Research Mine located near Pittsburgh, PA. However, this mine is technically not abandoned and therefore not subject to collapse and deterioration. On May 30, 2003 Groundhog finally entered an inaccessible abandoned mine in fully autonomous mode. The mine is known as the Mathies mine and is located in the same geographic area as the other mines. The core of this surface-accessible mine consists of two 1.5-kilometer long corridors which branches into numerous side corridors, and which are accessible at both ends.

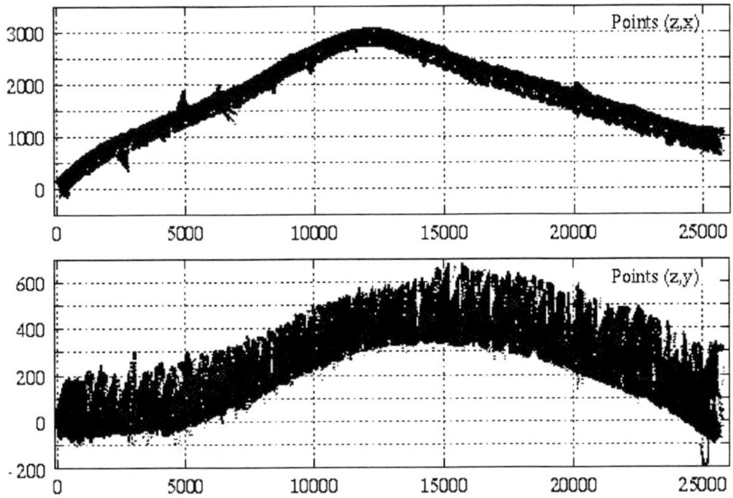

Fig. 6.43 Complete 3D map of the mine. Top: zx-map (view from top). Bottom: yx-map (view from side). z is the depth axis and and y the elevation. units: cm.

This was an important feature of this mine, as it provided natural ventilation and thereby reduced the chances of encountering combustible gases inside the mine.

To acquire an accurate 3D map of one of the main corridors, the robot was programmed to autonomously navigate through the corridor. 250 meters into the mine, the robot encountered a broken ceiling bar draping diagonally across its path. The robot made the correct decision to retract. The data acquired on these runs has provided us with models of unprecedented detail and accuracy, of subterranean spaces that may forever remain off limits for people.

Figure 6.42 presents two 3D scans and their alignment. Figure 6.43 shows the final result of the Mathies Mine mapping. The top plot shows the 2D map, i.e., xz-map, where z is the depth axis. The bottom part shows the elevation, i.e., the xy-map. The Groundhog robot had to overbear a height of 4 meters during its 250 meter long autonomous drive.

Mapping the Kvarntorp Mine

In a further experiment the mapping algorithms were tested in an underground mine, namely in the Kvarntorp mine outside of Örebro in Sweden. This mine is no longer in production, but was once used to mine limestone. Figure 6.44 shows a typical scene from one of the tunnels. The mine consists of around 40 km of tunnels, all in approximately one plane.

First of all, Kurt3D's driving capabilities have been tested. It turned out, that Kurt3D was able to drive inside this mine and the odometry produced

Fig. 6.44 Left: One of the tunnels in the Kvarntorp mine. Right: Kurt3D.

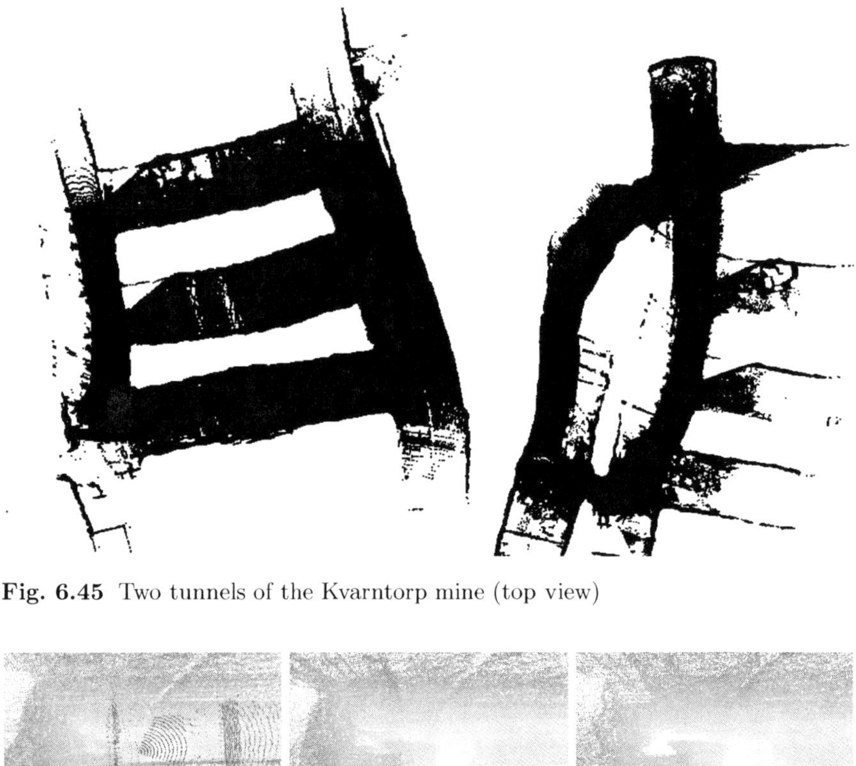

Fig. 6.45 Two tunnels of the Kvarntorp mine (top view)

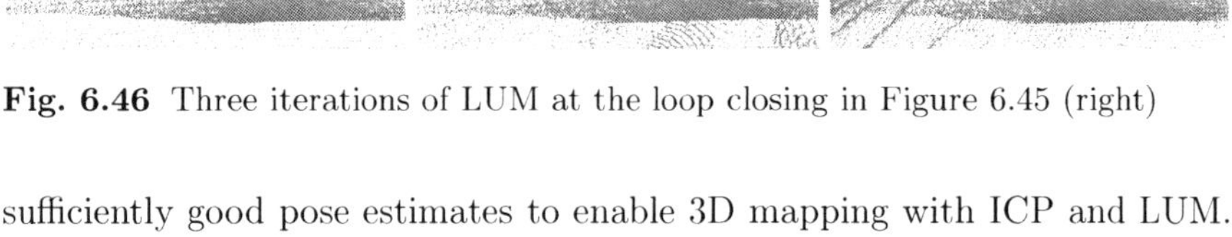

Fig. 6.46 Three iterations of LUM at the loop closing in Figure 6.45 (right)

sufficiently good pose estimates to enable 3D mapping with ICP and LUM. Several tunnels including cycles have been mapped. Figure 6.45 and 6.46 present some quantitative results.

6.5.2 Applications in Medicine

Holographic Recording and Automatic 3D Scan Matching of Living Human Faces

To treat diseases, injuries and congenital or acquired deformities of the head and neck, maxillo-facial surgeons deal with complex surgery. For example, the correction of disfiguring facial birth defects requires the manipulation of scull bones with maximum precision. The pre-operative simulation of such procedures requires a 3D computer model of the patient's face. We describe an approach to create such a 3D patient model by ultra-fast holographic recording and automatic scan matching of synchronously captured holograms. The pulsed hologram records the patient's portrait within a single laser shot (pulse duration approximately unit[35]ns). This so-called master-hologram contains the complete 3D spatial information which, due to the extremely short recording time, is not affected by involuntary patient movements.

In a second step, the real image of the hologram is optically reconstructed with a cw-laser. By moving a diffusor-screen through the real image, a series of 2D images are projected and digitized with a CCD camera. This process is referred to as hologram tomography [47]. Each projection shows the surface contour of the patient where the image is in focus. The method was first introduced as the *locus of focus* technique [114] in the context of non-medical imaging. Besides the desired intensity from in-focus points from the object contour, each captured image also contains a blurred background of defocused parts of the real image. The main problem of locating the surface is therefore to distinguish between focused and unfocused regions in each slice. This procedure yields a relief map of the visible (as seen from the hologram) parts of the patient. In order to record a complete 360° model of a patient, multiple holograms are recorded synchronously, i.e., with the same laser pulse.

Subsequently, the resulting relief maps are registered (i.e. their relative orientation is calculated) by automated scan matching. Given that the surface shapes are acquired independently from locus-of-focus analysis of two

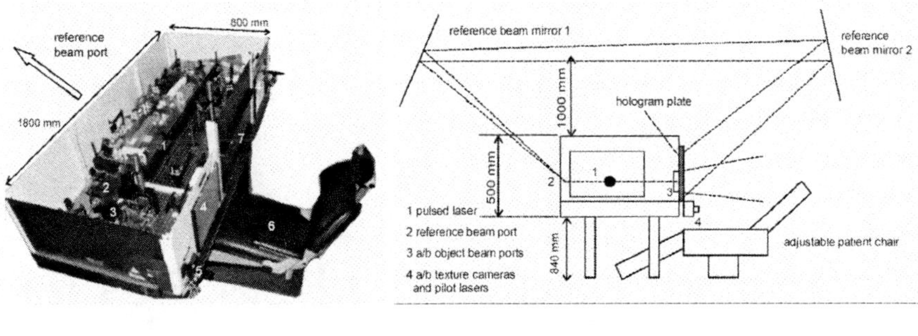

Fig. 6.47 Experimental acqusition of holograms. Left: Apparatus for face profiles. Right: Schematic illustration. Figure courtesy of D. Giel [47].

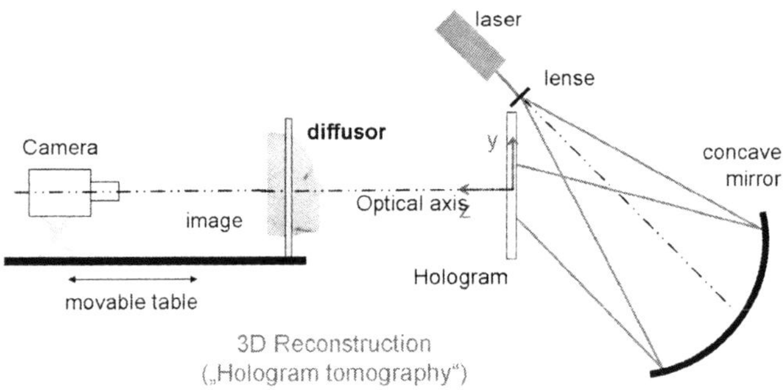

Fig. 6.48 As hologramm tomography the process of scanning a hologram using a diffusor is called. Afterwards depth images are calculated from camera images using depth from focus. Figure courtesy of D. Giel [47].

synchronously recorded holograms, such an approach yields 360° models of complex surfaces.

Hologram Tomography. Portrait holograms are recorded with a holographic camera of the type GP-2J (brand Geola) with master-oscillator and second harmonic generation. The resulting wavelength of 526.5 nm has a small penetration depth into skin to minimize light diffusion. Careful mode selection leads to a coherence length of approximately 6 m. The laser output is split into three beams: Two of them serve for homogeneous illumination of the object. They are expanded by concave lenses and diffusor plates at the output ports of the laser. The third beam serves as reference beam. The hologram plate (30 cm × 40 cm, VRP-M emulsion by Slavich) is developed with SM-6 and bleached with PBU-Amidol to obtain phase holograms. A frequency doubled cw Nd:YAG laser (COHERENT Verdi V-2) is used to reconstruct the holographic real image. To obtain the 2D projections of the real

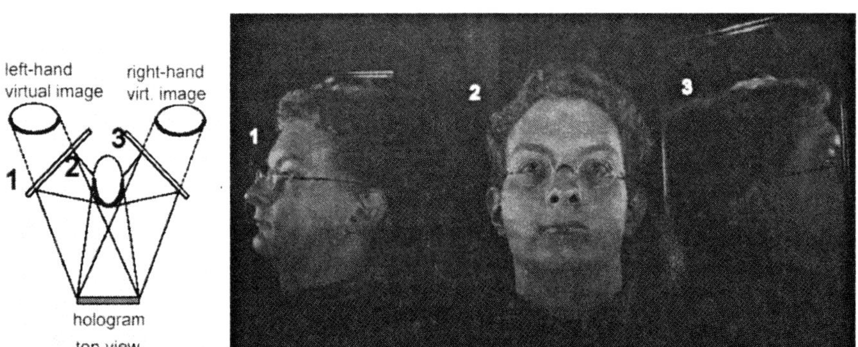

Fig. 6.49 Virtual image of a portrait hologram. The two side views are captured by mirrors. Figure courtesy of D. Giel [47].

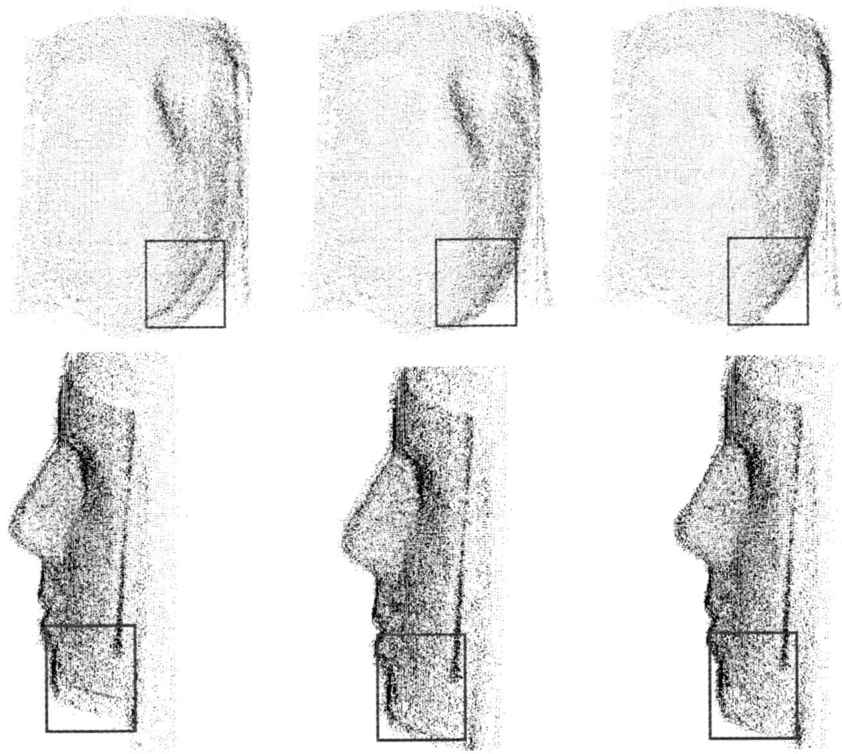

Fig. 6.50 Registration of two 3D reliefs with the ICP algorithms. Left: Initial alignment. Middle: Alignment after 4 iterations. Right: Final alignment after 85 iterations.

image, a diffusor (diffusor thickness 40 μm, diameter 380 mm) is moved on a computer controlled linear positioning stage (PI M-531.DD, max. resolution 10 μm) through the image volume. The diffusor images are digitized by a KODAK Megaplus ES 4.0 digital camera with 2048 x 2048 pixels.

To analyze the sequence of 2D-projections, the so-called slices, we use digital image processing. As an approximation we assume that the surface derived from a single hologram has no undercuts. Therefore there can be no two surface points with the same (x, y)-coordinate and the surface can be represented by a relief map. As already mentioned, each captured slice contains the specific focused information representing the object shape contour and a defocused background. The task is thus to distinguish between focused and defocused image regions: To evaluate the sharpness of an image in an conventional imaging system, several algorithms have been proposed in the field of image processing. We found that the best measure for image sharpness is the statistical variance $V_{(x,y)}$ of the light intensity on pixel adjacent to (x, y). For each lateral coordinate (x, y), the sharpness measure $V_{(x,y)}(z)$

is a positive, real number. The axial coordinate $z_{(x,y)}$ is assigned by choosing $z_{(x,y)}$ to satisfy

$$V_{(x,y)}(z_{(x,y)}) \geq V_{(x,y)}(z) \quad \forall\ z.$$

Thus each holographic real image gives a relief map of the object surface.

Figure 6.50 shows three frames of automatic alignment of 3D face scans.

Chapter 7
3D Maps with Semantics

Understanding environments is essential if a robotic system shall operate autonomously. This chapter proposes a complete technical cognitive system, consisting of a mobile robot, a 3D laser scanner and a set of algorithms for semantic environment mapping. We define a semantic 3D map as follows:

> A *semantic 3D map* for mobile robots is a metric map, that contains in addition to geometric information of 3D data points, assignments of these points to known structures or object classes.

Our approach uses 3D laser range and reflectance data on an autonomous mobile robot to perceive 3D objects. Starting from an empty map, several 3D scans acquired by the mobile robot Kurt3D in a stop-scan-go fashion, are merged in a global coordinate system by 6D SLAM as discribed in the previous parts of this book. Then, the coarse structure of the resulting 3D scene is interpreted using plane extraction and labeling exploiting background knowledge stored in a semantic net [89]. Afterwards, the 3D range and reflectance data are transformed into images by off-screen rendering. A cascade of classifiers, i.e., a linear decision tree, is used to detect and localize the objects [88]. Finally, the semantic map is presented using computer graphics. Figure 7.1 presents a system overview.

7.1 Scene Interpretation

7.1.1 Feature Extraction

A common technique for plane extraction is the region growing based approach, e.g., used by Hähnel et al. [54]. Starting from an initial mesh, neighboured planar triangles are merged iteratively. The drawback of this approach is the high computational demand.

Another well known algorithm for feature extraction from data sets is the RANSAC algorithm [1], used by Cantzler et al. [16]. RANSAC (Random Sample Consensus) is a simple algorithm for robust fitting of models in the presence of many data outliers. RANSAC first selects N data items randomly and uses them to estimate the parameters of the plane. The next

A. Nüchter: 3D Robotic Map.: The Simultaneous Local. & Map. Prob., STAR 52, pp. 155–172.
springerlink.com

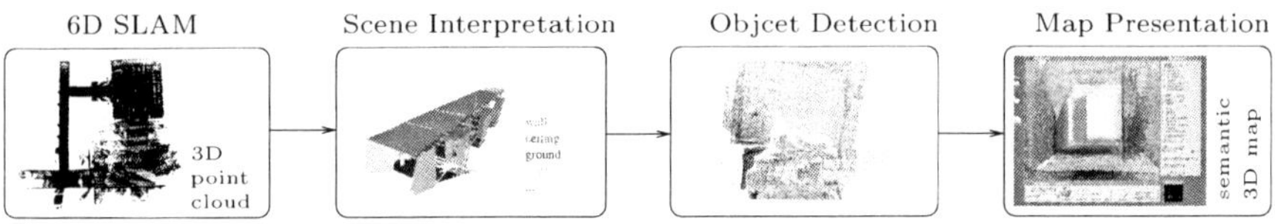

Fig. 7.1 3D maps with semantics: From left to right. Acquisition of a point cloud that consists of several 3D scans using 6D SLAM. Scene interpretation labels the basic elements in the scene. Object detection locates previously learned objects with 6 DoF. Finally, the semantic map is presented to the user.

step computes the number of data points fitting the model based on a user given tolerance. RANSAC accepts the fit, if the computed number exceeds a certain limit. Otherwise the algorithm iterates with other points [1].

Our algorithm is a mixture of the RANSAC and the ICP algorithm, and provides fast plane extraction for a point cloud. No prior meshing algorithms need to be applied. A plane p is defined by three 3D points ($\boldsymbol{p}_1, \boldsymbol{p}_2, \boldsymbol{p}_3 \in \mathbb{R}^3$) or by one 3D point and the surface normal ($\boldsymbol{p}_1$, $\boldsymbol{n}$ with $||\boldsymbol{n}|| = 1$, $\boldsymbol{p}_1, \boldsymbol{n} \in \mathbb{R}^3$). To detect a surface, the algorithm randomly selects a point and estimates a plane through two neighbouring data points. Now the data points $\boldsymbol{x} \in \mathbb{R}^3$ are calculated that fulfil:

$$|(\boldsymbol{x} - \boldsymbol{p}_1) \cdot \boldsymbol{n}| < \varepsilon. \tag{7.1}$$

If this set of points exceeds a limit, e.g., 50 points, an ICP based optimization is started. All data points satisfying (7.1) form the model set M and are projected to the plane to form the data set D for each iteration of the ICP algorithm. Minimizing the ICP error function (4.1) by transforming the plane with this point-to-plane metric takes only a few iterations. The time consuming search is replaced by direct calculation of the closest point and the transformation $(\boldsymbol{R}, \boldsymbol{t})$ is efficiently calculated [61]. Given the best fit, all plane points are marked and subtracted from the original data set. The algorithm terminates after all points have been tested according to Eq. (7.1).

The extracted 3D planes are unbounded in size. Surfaces are finally extracted from the points by mapping them onto the planes. A quadtree based method generates the surfaces. Figure 7.3 shows an example with 7 extracted planes of a single 3D scan containing 58680 range data points.

7.1.2 Interpretation

The scene interpretation uses the features, i.e., planes previously computed. The background for interpretation comprises generic architectural knowledge. A model of an indoor scene is implemented as a semantic net based on the idea of Grau et al. [48] that is also used by Cantzler et al. [16].

Nodes of a semantic net represent entities of the world. The relationships between the entities are encoded using different connections. Possible

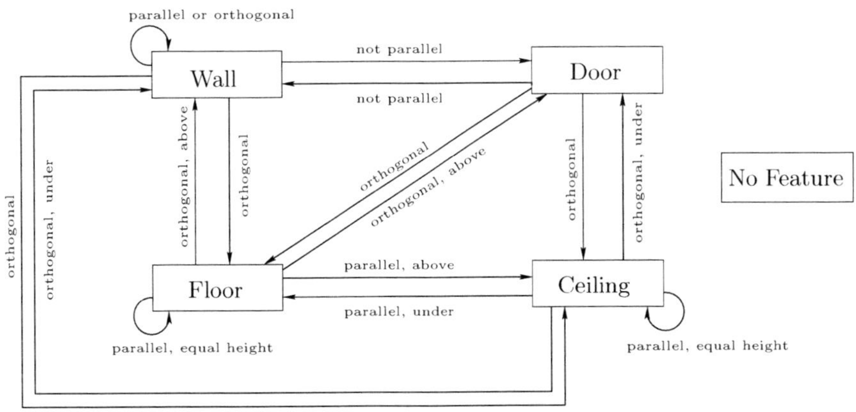

Fig. 7.2 Semantic net for coarse scene interpretation

labels of the nodes are L = {`Wall`, `Floor`, `Ceiling`, `Door`, `No Feature`}. The relationships between the features are R = {`parallel`, `orthogonal`, `above`, `under`, `equalheight`}. The labels `above` and `under` are relative to their plane and hence not commutative. Figure 7.2 shows the entities and their relations. The entity `door` represents an *open* door. The semantic net can be easily extended. Further entities have to be accompanied by a more sophisticated feature detection. Here we concentrate on plane detection so that the semantic net is a subset of all indoor environments.

Prolog is used to implement and externalize the semantic net. The net is encoded by definite Horn clauses. The nodes of the net are arguments and the arcs define relations on the nodes. All facts for the relation `parallel` are shown. For encoding the label `nofeature`, a condition is used. This prevents Prolog's unification algorithm from assigning planes with the label `nofeature`. Next, the Prolog implementation of the semantic net is given:

```
parallel(floor,floor).
parallel(ceiling,floor).
parallel(ceiling,ceiling).
parallel(floor,ceiling).
parallel(wall,wall).
parallel(X,_) :- X == nofeature.
parallel(_,X) :- X == nofeature.
orthogonal(ceiling,door).
orthogonal(ceiling,wall).
...
equalheight(floor,floor).
equalheight(ceiling,ceiling).
equalheight(door,_).
...
```

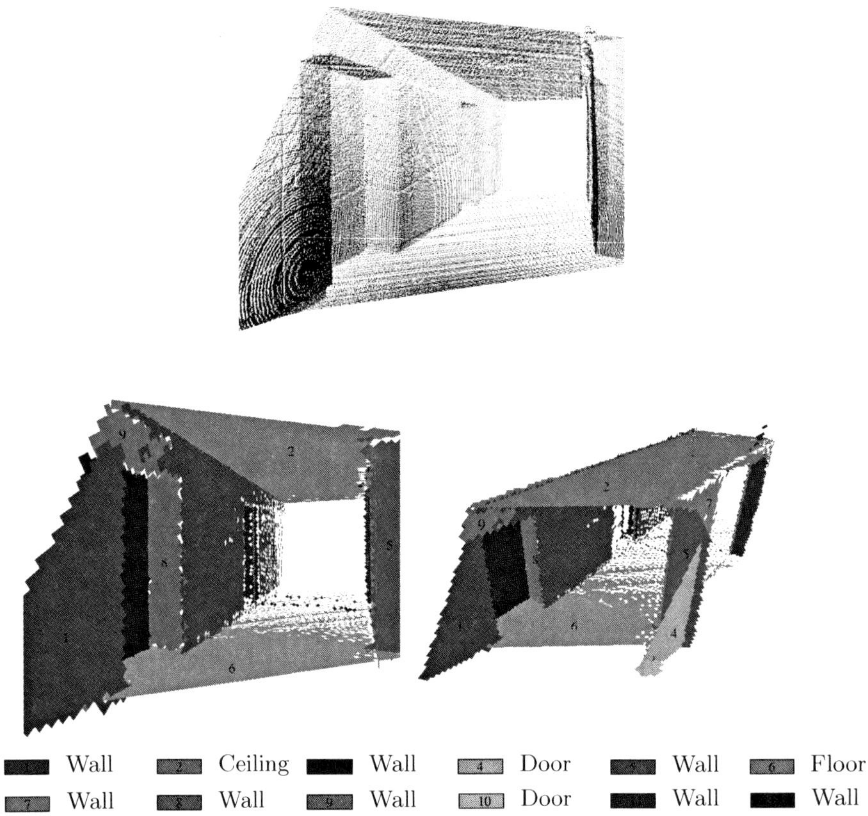

Fig. 7.3 Left: Point cloud. Middle and right: Extracted planes and semantic interpretation.

In addition to the representation of the semantic net, a clause of the following form is compiled from the analysis of the planes. The planes are represented by variables P0, P1, etc.:

```
labeling(P0,P1,P2,P3,P4) :-
          parallel(P0,P1),under(P0,P1),
          orthogonal(P0,P2),under(P0,P2),
          orthogonal(P0,P3),under(P0,P3),
          parallel(P0,P4),above(P0,P4),
...
```

Prolog's unification algorithm is used to assign a consistent labeling to the planes:

```
consistent_labeling(P0,P1,P2,P3,P4) :-
            labeling(P0,P1,P2,P3,P4).
```

The label `nofeature` is considered, iff the unification fails. In this case, an additional Horn clause is used to generate a consistent labeling with explicit unification of one variable with `nofeature`. All combinations are computed to unify the variable:

```
consistent_labeling(P0,P1,P2,P3,P4) :-
           comb([P0,P1,P2,P3,P4],[nofeature]),
           labeling(P0,P1,P2,P3,P4).
```

The process is continued with assigning two variables the label to `nofeature`, and so on until a Prolog's unification succeeds:

```
consistent_labeling(P0,P1,P2,P3,P4) :-
           comb([P0,P1,P2,P3,P4],[nofeature,nofeature]),
           labeling(P0,P1,P2,P3,P4).
```

The order of the rules above is important, as in all Prolog programs. Prolog searches for clauses from top to bottom. The following three clauses are used to compute all combinations:

```
comb(_,[]).

comb([X|T],[X|Comb]) :- comb(T,Comb).

comb([_|T],[X|Comb]) :- comb(T,[X|Comb]).
```

Finally the following query is submitted:

```
consistent_labeling(P0,P1,P2,P3,P4).
```

and the automaticly generated Prolog program starts and computes the solution. Table 7.1 shows the computation time (Pentium IV-2400, SWI-Prolog [101]) of the Prolog program and a previous complete depth first search implementation [89].

Table 7.1 Computation time for matching the semantic net with the planes

number of planes	backtracking	Prolog
5	93.51 ms	89.33 ms
7	155.14 ms	101.81 ms
13	589.11 ms	313.79 ms

7.1.3 Application: Model Refinement

In this section an application of the semantic interpretation is presented. We use the specific knowledge that has been generated from the acquired data and generic knowledge, to correct the sensor data.

Due to imprecise measurements or registration errors, the 3D data might be erroneous. These errors lead to inaccurate 3D models. The semantic interpretation enables us to refine the model. The planes are adjusted such that they explain the 3D data, and the semantic constraints like parallelism or orthogonality are enforced.

To enforce the semantic constraints, the model is first simplified. A preprocessing step merges neighbouring planes with equal labels, e.g., two ceiling planes. This simplification process increases the point-to-plane distance, which has to be reduced in the following main optimization process. This optimization uses an error function to enforce the parallelism or orthogonality constraints. The error function consists of two parts. The first part accumulates the point to plane distances and the second part accumulates the angle differences given through the constraints. The error function has the following form:

$$E(P) = \sum_{p_i \in P} \sum_{\boldsymbol{x} \in p_1} ||(\boldsymbol{x} - \boldsymbol{p}_{i1}) \cdot \boldsymbol{n}_i|| + \gamma \sum_{p_i \in P} \sum_{p_j \in P} c_{i,j}, \tag{7.2}$$

where $c_{i,j}$ expresses the parallelism or orthogonality constraints, respectively, according to

$$\begin{aligned} c_{i,j} &= \min\{|\arccos(\boldsymbol{n}_i \cdot \boldsymbol{n}_j)|, |\pi - \arccos(\boldsymbol{n}_i \cdot \boldsymbol{n}_j)|\}, \quad \text{and} \\ c_{i,j} &= |\tfrac{\pi}{2} - \arccos(\boldsymbol{n}_i \cdot \boldsymbol{n}_j)|, \quad \text{respectively.} \end{aligned}$$

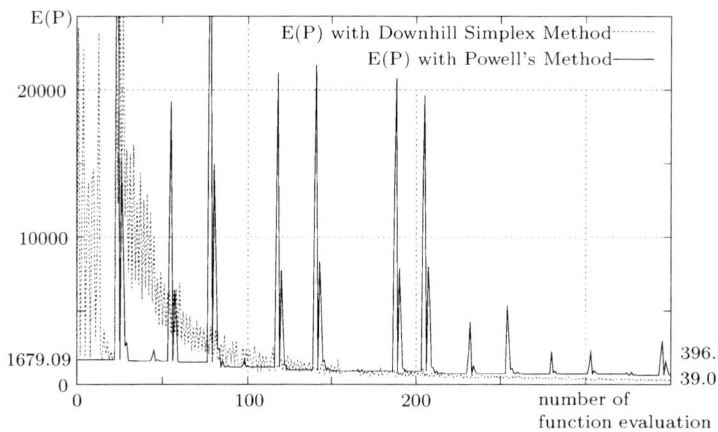

Fig. 7.4 Minimization of the error function starting with $E(P) = 1679.09$. Powell's method finds a minimum at 396.5 and the downhill simplex reaches a minimum at 39.0.

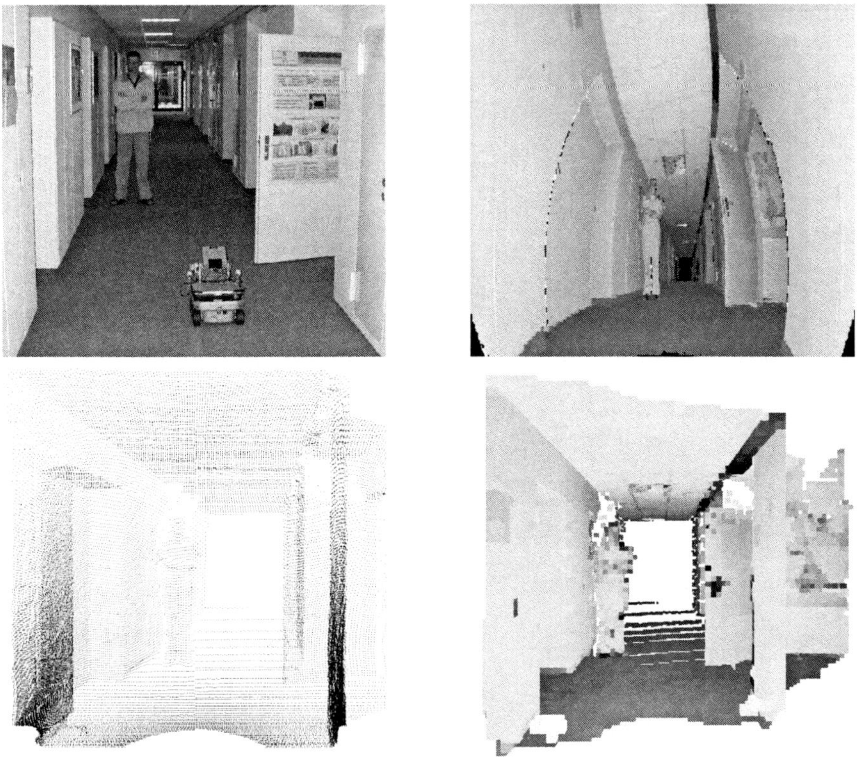

Fig. 7.5 From top left to bottom right: Photo of the scanned scene; Reflectance values acquired by the 3D laser range finder (distorted, since the rotation of the scanner is not considered [118]); 3D point cloud; rendered scene with reflectance values; continued as Figure 7.6

Minimization of $E(\mathbf{P})$ (Eq. (7.2)) is a nonlinear optimization process.

The time consumed for optimizing $E(\mathbf{P})$ increases with the number of plane parameters. To speed up the process, the normal vectors $\boldsymbol{n}$ of the planes are specified by spherical coordinates, i.e., two angles α, β. The point $\boldsymbol{p}_1$ of a plane is reduced to a fixed vector pointing from the origin of the coordinate system in the direction of $\boldsymbol{p}_1$ and its distance d. The minimal description of all planes $\mathbf{P}$ consists of the concatenation of $\boldsymbol{p}_i$, with $\boldsymbol{p}_i = (\alpha_i, \beta_i, d_i)$, i.e., a plane $\boldsymbol{p}_i$ is defined by two angles and a distance.

Two methods are applicable for minimization of Eq. (7.2), namely Powell's method and the downhill simplex method (see Appendix A.6). Powell's method is suitable, since the optimal solution is close to the starting point. Powell's method finds search directions with a small number of error function evaluations of eq. (7.2). Gradient descent algorithms have difficulties, since no derivatives are available. Similarily the downhill simplex method does not need derivatives.

Experimental evaluations for the environment test settings show that the minimization algorithm finds a local minimum of the error function (7.2). The computed description of the planes fits the data and the semantic model. Figure 7.4 shows the minimization of Eq. (7.2) with the downhill simplex method in comparison with Powell's method. The downhill simplex method performs worse during the first steps, but reaches a better minimum than Powell's method. The peaks in $E(\mathbf{P})$ produced by Powell's method are the result of search directions $\boldsymbol{i}$ cross a "valley" in combination with large steps in Brent's line minimization algorithm (cf. Appendix A.6).

Final Refinement and Results

The semantic description, i.e., the ceiling and walls enables to transform the orientation of the model along the coordinate axes. Therefore it is not necessary to transform the model interactively into a global coordinate system or to stay in the coordinates given by the first 3D scan.

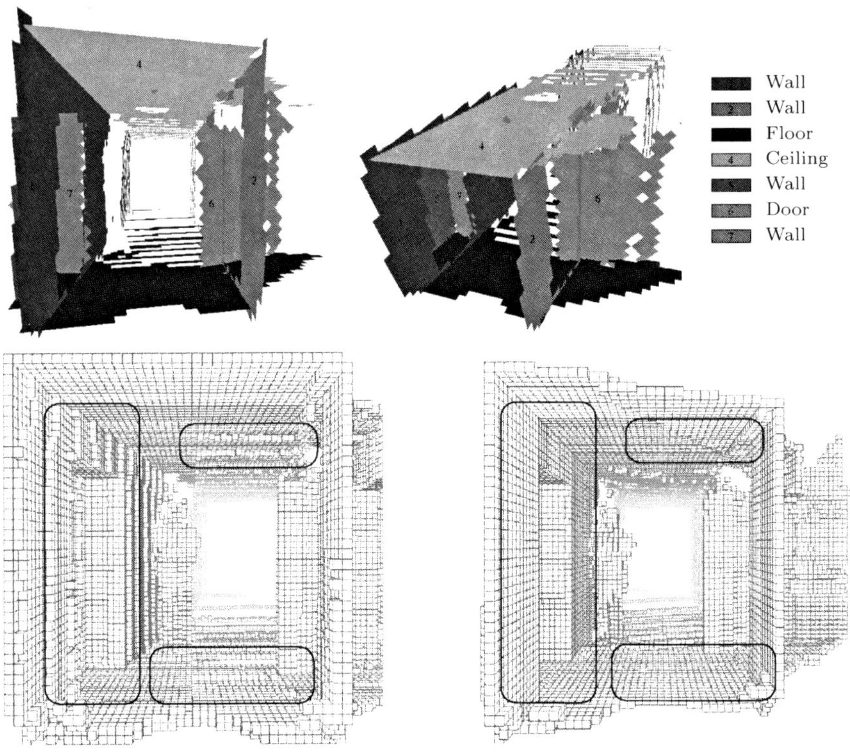

Fig. 7.6 Top: Extracted planes with semantic interpretation. Bottom: Unconstrained mesh and constrained mesh (cf. Figure 7.5).

The proposed methods have been tested in several experiments with the autonomous mobile robot Kurt3D in the GMD Robobench, a standard office environment for the evaluation of autonomous mobile robots. Figure 7.3 and 7.5 shows an example 3D point cloud (single 3D scan with 58680 points) and the semantic interpretation. The figure shows the reduction of the jitters at the floor, ceiling and walls (circled). The orientation of the model in the bottom image is transformed along the axes of the coordinate system and the meshing algorithm produces flat walls. Hereby an octree-based algorithm [103] generates the mesh (cube width: 5 cm). The total computation time for the complete optimization is about 2.4 seconds (Pentium-IV-2400).

7.2 Object Localization in 3D Data

Five steps are necessary to reliably localize an object, that is stored in a database, within the 3D laser range data, namely image rendering, object detection using a cascade of classifiers, ray tracing, model fitting and evaluation. Figure 7.7 visualizes these steps, that are described next.

7.2.1 Object Detection

Rendering Images from Scan Data

After scanning, the 3D data points are projected by an off-screen `OpenGL`-based rendering module onto an image plane to create 2D images. The camera for this projection is located in the laser source. Figure 7.12 shows reflectance images and rendered depth images (distances encoded by gray-values) as well as point clouds.

Feature Detection using Integral Images

There are many motivations for using features rather than pixels directly. For mobile robots, a critical motivation is that feature-based systems operate much faster than pixel-based systems [130]. The features used here have

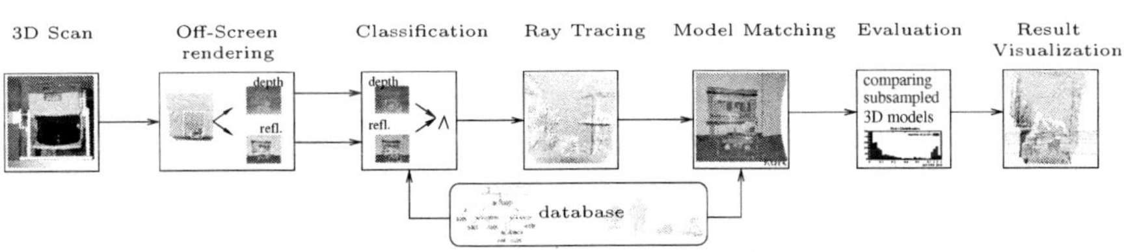

Fig. 7.7 After acquiring 3D scans, depth and reflection images are generated. In these images, objects are detected using a learned representation from a database. Ray tracing selects the points corresponding to the 2D projection of the object. A 3D model is matched into these points, followed by an evaluation step.

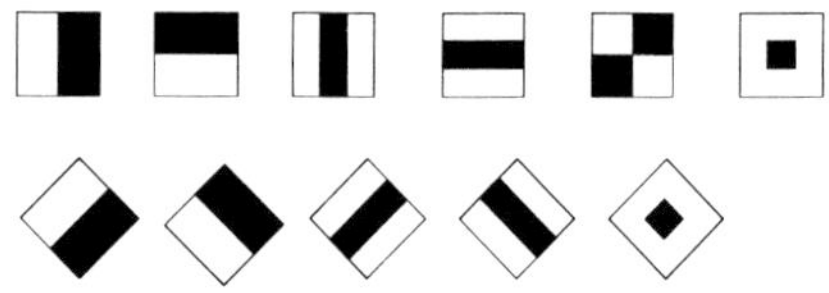

Fig. 7.8 Edge, line, diagonal and center surround features are used for classification

the same structure as the Haar basis functions, i.e., step functions introduced by Alfred Haar to define wavelets [53]. They are also used in [74,94,130]. Figure 7.8 shows the eleven basis features, i.e., edge, line, diagonal and center-surround features. The base resolution of the object detector is for instance 30×30 pixels, thus, the set of possible features in this area is very large (642592 features, see [74] for calculation details). In contrast to the Haar basis functions, the set of rectangle features is not minimal. A single feature is effectively computed on input images using integral images [130], also known as summed area tables [74]. An integral image I is an intermediate representation for the image and contains the sum of gray scale pixel values of image N with height y and width x, i.e.,

$$I(x,y) = \sum_{x'=0}^{x} \sum_{y'=0}^{y} N(x',y').$$

The integral image is computed recursively, using the formulas: $I(x,y) = I(x,y-1) + I(x-1,y) + N(x,y) - I(x-1,y-1)$ with $I(-1,y) = I(x,-1) = I(-1,-1) = 0$, therefore requiring only one scan over the input data. This intermediate representation $I(x,y)$ allows the computation of a rectangle feature value at (x,y) with height and width (h,w) using four references:

$$F(x,y,h,w) = I(x,y) + I(x+w,y+h) - I(x,y+h) - I(x+w,y).$$

For computing the rotated features, Lienhart et. al. introduced rotated summed area tables that contain the sum of the pixels of the rectangle rotated by 45° with the bottom-most corner at (x,y) and extending till the boundaries of the image [74]:

$$I_r(x,y) = \sum_{x'=0}^{x} \sum_{y'=0}^{x-|x'-y|} N(x',y').$$

The rotated integral image I_r is computed recursively, i.e., $I_r(x,y) = I_r(x-1,y-1) + I_r(x+1,y-1) + -I_r(x,y-1) + N(x,y) + N(x,y-1)$ using the start values $I_r(-1,y) = I_r(x,-1) = I_r(x,-2) = I_r(-1,-1) = I_r(-1,-2) = 0$. Four table lookups are required to compute the pixel sum of any rotated rectangle with the formula:

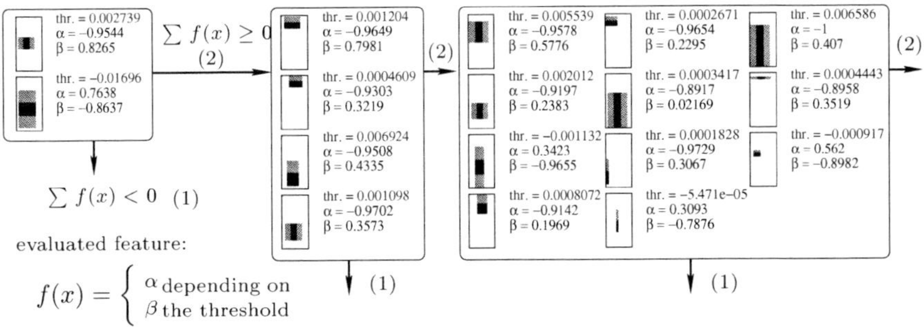

Fig. 7.9 The first three stages of a cascade of classifiers to detect an *office chair* in depth data. Every stage contains several simple classifiers that use Haar-like features with a threshold and returning values α, β for $h(x)$.

$$\begin{aligned} F_r(x, y, h, w) = {} & I_r(x + w - h, y + w + h - 1) + \\ & I_r(x, y - 1) - I_r(x - h, y + h - 1) - \\ & I_r(x + w, y + w - 1). \end{aligned}$$

Since the features are compositions of rectangles, they are computed with several lookups and subtractions weighted with the area of the black and white rectangles.

To detect a feature, a threshold is required. This threshold is automatically determined during a fitting process, such that a minimal number of examples is misclassified. Furthermore, the return values (α, β) of the feature are determined, such that the error on the examples is minimized. The examples are given in a set of images that are classified as positive or negative samples. The set is also used in the learning phase that is briefly described next.

Learning Classification Functions

The Gentle Ada Boost Algorithm is a variant of the powerful boosting leaning technique [39]. It is used to select a set of simple features to achieve a given detection and error rate. In the following, a detection is referred as hit and an error as a false alarm. The various Ada Boost algorithms differ in the update scheme of the weights. According to Lienhart et al. the Gentle Ada Boost Algorithm is the most successful learning procedure tested for face detection applications [73].

The learning is based on N weighted training examples $(x_1, y_1), \ldots, (x_N, y_N)$, where x_i are the images and $y_i \in \{-1, 1\}, i \in \{1, \ldots, N\}$ the classified output. At the beginning of the learning phase the weights w_i are initialized with $w_i = 1/N$. The following three steps are repeated to select simple features until a given detection rate d is reached:

1. Every simple classifier, i.e., a single feature, is fit to the data. Hereby the error e is calculated with respect to the weights w_i.

2. The best feature classifier h_t is chosen for the classification function. The counter t is increased.
3. The weights are updated with $w_i := w_i \cdot e^{-y_i h_t(x_i)}$ and renormalized.

The final output of the classifier is

$$\mathrm{sgn}(\sum_{t=1}^{T} h_t(x)),$$

where $h(x) = \alpha$, if $x \geq$ thr. and $h(x) = \beta$ otherwise. α and β are the output of the fitted simple feature classifiers, that depend on the assigned weights, the expected error and the classifier size. Next, a cascade based on these classifiers is built.

The Cascade of Classifiers

The performance of a single classifier is not suitable for object classification, since it produces a high hit rate, e.g., 0.999, but also a high error rate, e.g., 0.5. Nevertheless, the hit rate is significantly higher than the error rate. To construct an overall good classifier, several classifiers are arranged in a cascade, i.e., a degenerated decision tree. In every stage of the cascade, a decision is made whether the image contains the object or not. This computation reduces both rates. Since the hit rate is close to one, their multiplication results also in a value close to one, while the multiplication of the smaller error rates approaches zero. Furthermore, this speeds up the whole classification process, since large parts of the image do not contain relevant data. These areas can be discarded quickly in the first stages.

An overall effective cascade is learned by a simple iterative method. For every stage, the classification function $h(x)$ is learned until the required hit rate is reached (cf. Fig 7.9). The process continues with the next stage using the correctly classified positive and the currently misclassified negative examples. These negative examples are random image parts generated from the given negative examples that have passed the previous stages and thus are misclassified. This bootstrapping process is the most time consuming in the training phase. The number of simple classifiers used in each classifier may increase with additional stages. Figure 7.9 shows a learned cascade for the object *office chair* in depth data.

Application of the Cascades

Several experiments were made to evaluate the performance of the proposed approach with two different kinds of images, namely, reflectance and depth images. Both types are acquired by the 3D laser range finder and are practically light invariant. About 200 representation of the objects were taken in addition to a wide variety of negative examples without any target object.

Table 7.2 Object name, number of stages used for classification versus hit rate and the total number of false alarms using the single and combined cascades. The test sets consist of 89 images rendered from 20 3D scans. The average processing time is also given, including the rendering, classification, ray tracing, matching and evaluation time.

object	# of stages	detection rate (refl. img. vs. depth img.)	false alarms (refl. img. vs. depth img.)	average processing time
chair	15	0.767 (0.867 / 0.767)	12 (47 / 33)	1.9 sec
kurt robot	19	0.912 (0.912 / 0.947)	0 (5 / 7)	1.7 sec
volksbot robot	13	0.844 (0.844 / 0.851)	5 (42 / 23)	2.3 sec
human	8	0.961 (0.963 / 0.961)	1 (13 / 17)	1.6 sec

The detection starts with the smallest classifier size, e.g., 16×40 pixel for the human classifier, 23×30 for the volksbot classifier. The image is searched from top left to bottom right by applications of the cascade. To detect objects on larger scales, the detector is rescaled. An advantage of the Haar-like features is that they are easily scalable. Each feature requires only a fixed number of lookups in the integral image, independent of the scale. Time-consuming picture scales are not necessary to achieve scale invariance. Figure 7.12 show examples of the detection.

To decrease the false detection rate, we combine the cascades of the depth and reflectance images. There are two possible ways for combining: Either the two cascades run interleaved or serial and represent a logical "and" [88]. The joint cascade decreases the false detection rate close to zero. To avoid

Fig. 7.10 The object *printer* in front of two different backgrounds

Fig. 7.11 Object points estimation using ray tracing. Top left: All points inside a detection area are extracted. Top Right and bottom: 3D view. 3D points inside the detector area (viewing cone) are red colored.

the reduction of the hit rate, several different off-screen rendered images are used, where the virtual camera is rotated and the apex angle is changed [88].

Efficient Classifier Learning

Learning needs multiple positive and negative examples to construct a classifier. Negative examples are supplied by using gray scale images. To make learning efficient, we developed a strategy to generate several positive training images. After the object that has to be learned was scanned once, the 3D data points belonging to the object are manually extracted. Afterwards, different backgrounds, i.e., grayscale images, are placed behind the object using a 3D rendering tool. Finally the object is rotated in and 3D views are

Fig. 7.12 Examples of object detection and localization. From Left to right: (1) Detection using single cascade of classifiers. Dark gray: detection in reflection image, light gray: detection in depth image. (2) Detection using the combined cascade. (3) Superimposed to the depth image is the matched 3D model. (4) Detected object in the raw scanner data, i.e., point representation.

saved using off-screen drawing mode. Figure 7.10 shows the object *printer* in front of two different backgrounds. Using just one 3D scan to learn an object does not erode the quality of object detection.

7.2.2 Object Localization

Object Points Estimation

After object detection in a 2D projection the algorithm finds the corresponding 3D points using ray tracing. All 3D points that have been projected onto the classified area are retrieved using a special OpenGL projection matrix. Figure 7.11 (right) shows a rendering of raytraced 3D points.

Model Matching and Evaluation

After the 3D data (set $\mathcal{D}$) that contain the object is found, a given 3D model from the object database is matched into the point cloud. The model $\mathcal{M}$ is also saved as 3D point cloud in the database. The well-known iterative closest points algorithm (ICP) is used to find a matching [11] (cf. Eq. (4.1)). Afterwards the matching result is evaluated using normalized subsampling [87]. Figure 7.12 shows results of the object detection.

7.3 Semantic 3D Maps

In this section, we integrate the previously described algorithms into a complete system for semantic 3D mapping. Prior mapping, the object database needs to be initialized and filled with object descriptions. This is achieved by scanning the objects once and storing the object points and corresponding cascade.

The actual semantic mapping is done by the following steps (cf. Figure 7.1):

1. 3D mapping of the environment:
   a. Tele operated acquisition of 3D laser scans.
   b. Globally consistent scan registration using 6D SLAM.
2. Interpretation of the coarse 3D map structure. 3D surfaces are extracted and labeled using their orientation.
3. Extraction of objects from each 3D scan using classification in range and depth images and model matching. The 6D pose of the objects is estimated.
4. Visualization of the semantic map.

Due to the 3D nature of the semantic map, a rendering application is necessary for visualization. In our implementation the following information are accessible:

- Object information, i.e., the label of the selected object is visualized. Note: The label is attached in 3D to the 3D object, such that the label might be viewed from behind, e.g., Figure 7.13 two bottom rows. In addition the program prints the 6D pose of the object as $(x, y, z, \theta_x, \theta_y, \theta_z) \in \mathbb{R}^6$.
- All global 3D coordinates, i.e., $(x, y, z) \in \mathbb{R}^3$ of the scanned points.
- Distances between 3D points and to the current view pose.
- Information about the robot poses at which a 3D was acquired and trajectory data.
- Texture information, that can be acquired by the robots cameras.

Fig. 7.13 Semantic 3D maps of the AVZ building at University of Osnabrück. The rendered views show an office environment. Labels derived by coarse scene interpretation are black, whereas labels in gray are the result of object localization. Left: 3D point cloud with labeled objects. Middle: Coarse scene interpretation. Right: Reflectance values.

With the following experiment, we demonstrate exemplarily, how the previously described algorithm yields a semantic map. The experiment was carried out in an office environment. A corridor and some offices have been mapped. In the initialization phase we added 12 objects, like different chairs, printers, plants or even a human to the database. After 3D scanning of the test course, the 3D model contained 32 3D scans with about 2.5 million points in total. The overall map post-processing time was 4.5 minutes. 82% of all 3D points were assigned to a semantic label and since we had one object detection failure, i.e., one false detection of a fire extinguisher $\sim$0.1% of the labels wase wrong. Figure 7.13 presents a visualization of the results. An animation of the scene can be downloaded using the following link: `http://www.informatik.uni-osnabrueck.de/nuechter/videos.html`

The approach taken here has some limitations, e.g., the learning phase is separate of the semantic mapping phase. Furthermore, the computing time object detection grows linearly with the number of objects in the database. To reduce this effect, we proposed a saliency-based enhancement. In [41] the detector, i.e., the application of the cascades (cf. Sec. 7.2.1) is focused on salient image regions to reduce computational requirements.

Chapter 8
Conclusions and Outlook

2D laser range finders are a de facto standard in robotics and metrical planar 2D mapping of well-defined indoor environments is considered solved. However, mapping unstructured environments remains challenging. Capturing environments in 3D enables robots to cope with these unstructured environments. A representation of the robot's pose with 6 DoF further enables the mapping vehicle to drive on non-flat ground.

This book has presented algorithms for 3D mapping. Input are 3D data, i.e., range data. The output are transformations of the poses of the sensor, resp. robot. These transformations are rigid and consist of a rotation and a translation. When applied to the robot pose, the pose is corrected and the sensor readings can then be registered in a globally consistent map. Hence, the presented algorithms solve the simultaneous localization and mapping problem with 6 DoF for 3D maps. The technical basis of the algorithms is a fast and reliable scan matching implemented through closest point computation. This book discusses implementations, running times and their lower bounds.

The presented algorithms have been implemented with performance in mind. A detailed evaluation and several mapping examples round off the book in Chapter 6. The ICP algorithm for reasonable sized point clouds runs on-line on-board of the mobile robot Kurt3D. Using this technology it is possible to digitalize large areas, like urban environments. Continuously rotating 3D scanner, like the one of the Lebniz Universität Hannover presented in Section 6.3 allows 3D scanning and driving at the same time. With this technology it is possible to acquire 3D point clouds in real time. Possible applications are city and street mapping. Several 3D point clouds acquired at different days permit even removing artefacts like humans walking by or parking cars.

The last chapter combines semantics with 3D maps. Semantic interpretation is added in two steps: First, a semantic net is used for coarse scene interpretation of indoor scenes by labeling planes. Second, previously learned

A. Nüchter: 3D Robotic Map.: The Simultaneous Local. & Map. Prob., STAR 52, pp. 173–175.
springerlink.com

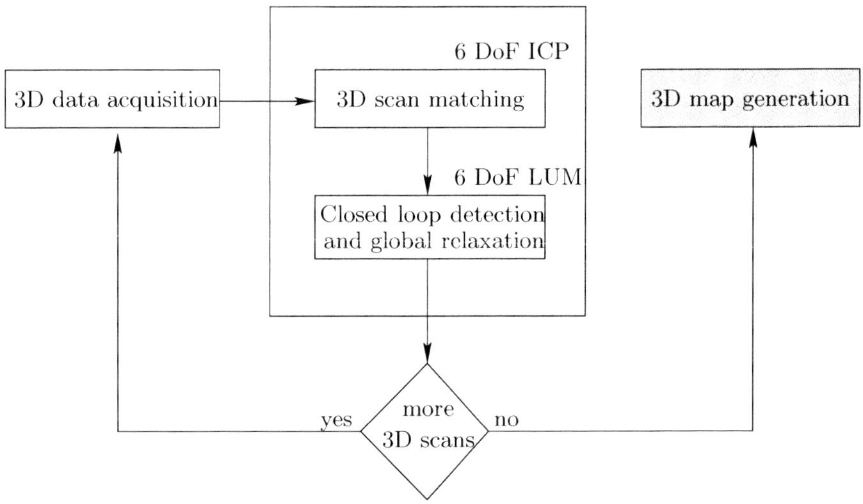

Fig. 8.1 Future work: After solving SLAM a polygonal map has to be generated

objects are detected. All these algorithms also make use of 6 DoF, e.g. the pose of every object is also specified in 6 DoF.

Future work on the robot Kurt3D concentrates on the following aspects:

Obect detection. A silhouette-based object detection has been included in the robot control software. Stiene shows that contours of objects standing on the ground can be easily extracted in depth data, if the semantic "ground" of the points is known [115]. Based on this fact a contour classifier has been implemented.

Planning. The robot control architecture of Kurt3D is extended by a meaningful exploration algorithm. The exploration must regard the 3D sensor data and the 6D SLAM algorithm as well as the fact, that the robot is able to control only 3 DoF, that is the robot moves on a surface.

Rescue robotics. The mobile robot Kurt3D cannot drive in challenging areas, due to its driving capabilities. NIST is currently fostering research on highly mobile robot platforms. This is done by the introduction of so-called step fields in the RoboCup rescue competition. Wooden quads with a cross section of 10 cm×10 cm form the step fields and simulate brick piles that are often found in earthquake areas.
In future work we will show, how the presented algorithm performs on highly mobile platforms. In this context cooperative construction of 3D maps by several robots is also subject to research.

Simulation. The development of algorithms for autonomous mobile robots is time consuming, due to the large number of experiments with real systems. Simulations of robots help reducing these efforts. In future work we will connect USARSIM to the Kurt3D robot control architecture. Here the instance will be UOSSIM a simulation of the office environment at the

University of Osnabrück [58]. USARSim (Unified System for Automation and Robot Simulation) is a high-fidelity simulation of robots and environments based on the Unreal Tournament game engine. It is intended as a research tool and is the basis for the RoboCup rescue virtual robot competition.

Human-robot interaction. The Human-robot interaction (HRI) for mobile robots is also an upcoming issue, since the current state of the art are interactions of the type tele-operation. The operator determines the path of the vehicle, based on video and sensor-feedback. Therefore such a robot can often only be handled by an expert, usually the programmer of the interface. Since there is the need for non-experts to control robots, for instance in the rescue robotic context, the human-robot interaction is an important issue [110].

Future work in the 3D mapping context will concentrate on post-processing the 3D data (see Figure 8.1). The output of the 6D SLAM algorithm is a 3D point cloud that has to be post-processed to compute a polygonal model. Surfaces are commonly represented by polygonal meshes, particularly triangle meshes, a standard data structure in computer graphics to represent 3D objects.

Appendix A
Math Appendix

A.1 Representing Rotations

A.1.1 Euler Angles

Besides the three Cartesian coordinates x, y and z describing the position, the three angles θ_x, θ_y and θ_z describe the orientation in $\mathbb{R}^3$. Position and orientation are referred as pose. The orientation θ_x, θ_y and θ_z is the rotation about the principal axes, i.e., $(1,0,0)$, $(0,1,0)$ and $(0,0,1)$. The rotation matrices are given as follows:

$$\boldsymbol{R}_x = \begin{pmatrix} 1 & 0 & 0 \\ 0 & \cos\theta_x & -\sin\theta_x \\ 0 & \sin\theta_x & \cos\theta_x \end{pmatrix}$$

$$\boldsymbol{R}_y = \begin{pmatrix} \cos\theta_y & 0 & \sin\theta_y \\ 0 & 1 & \\ -\sin\theta_y & 0 & \cos\theta_y \end{pmatrix}$$

$$\boldsymbol{R}_z = \begin{pmatrix} \cos\theta_y & -\sin\theta_y & 0 \\ \sin\theta_y & \cos\theta_y & 0 \\ 0 & 0 & 1 \end{pmatrix}.$$

The overall rotation matrix is computed as $\boldsymbol{R} = \boldsymbol{R}_{\theta_x,\theta_y,\theta_z} = \boldsymbol{R}_x\boldsymbol{R}_y\boldsymbol{R}_z$, i.e.,

$$\boldsymbol{R} = \begin{pmatrix} \cos\theta_y\cos\theta_z & -\cos\theta_y\sin\theta_z & \sin\theta_y \\ \cos\theta_z\sin\theta_x\sin\theta_y + \cos\theta_x\sin\theta_z & \cos\theta_x\cos\theta_z - \sin\theta_x\sin\theta_y\sin\theta_z & -\cos\theta_y\sin\theta_x \\ \sin\theta_x\sin\theta_z - \cos\theta_x\cos\theta_z\sin\theta_y & \cos\theta_z\sin\theta_x + \cos\theta_x\sin\theta_y\sin\theta_z & \cos\theta_x\cos\theta_y \end{pmatrix}.$$

Note: The matrix above depends on the order of the multiplication. Different rotation matrices are achieved by $\boldsymbol{R} = \boldsymbol{R}_{\theta_z,\theta_y,\theta_x} = \boldsymbol{R}_z\boldsymbol{R}_y\boldsymbol{R}_x$ or $\boldsymbol{R} = \boldsymbol{R}_{\theta_y,\theta_x,\theta_z} = \boldsymbol{R}_y\boldsymbol{R}_x\boldsymbol{R}_z$, etc. Furthermore, note that a gimbal lock occurs when the axes of two of the three angles needed to compensate for rotations in three dimensional space are driven to the same direction.

A.1.2 Axis Angle

Rotations can be represented as a rotation axis and a rotation angle as given in Figure A.1. Given a unit vector $\boldsymbol{n} = (n_x, n_y, n_z)^T$ and an angle φ then the rotation of $\boldsymbol{p}$ to $\boldsymbol{p}' = \boldsymbol{q}$ is given by

$$\begin{aligned}\boldsymbol{p}' &= \overrightarrow{\boldsymbol{on}} + \overrightarrow{\boldsymbol{nv}} + \overrightarrow{\boldsymbol{vq}}\\ &= \boldsymbol{n}(\boldsymbol{n}\cdot\boldsymbol{r}) + (\boldsymbol{r} - \boldsymbol{n}(\boldsymbol{n}\cdot\boldsymbol{r}))\cos\theta + (\boldsymbol{r}\times\boldsymbol{n})\sin\theta\\ &= \boldsymbol{p}\cos\theta + \boldsymbol{n}(\boldsymbol{n}\cdot\boldsymbol{r})(1-\cos\theta) + (\boldsymbol{r}\times\boldsymbol{n})\sin\theta. \end{aligned} \tag{A.1}$$

The resulting term written as rotation matrix is

$$\boldsymbol{R}_{\boldsymbol{n},\theta} = \mathbb{1}_3 + (\boldsymbol{N})\sin\theta + (\boldsymbol{N})^2(1-\cos\theta)$$

where $\boldsymbol{N}$ is defined as follows

$$\boldsymbol{N} = \begin{pmatrix} 0 & -n_z & n_y \\ n_z & 0 & -n_x \\ -n_y & n_x & 0 \end{pmatrix}.$$

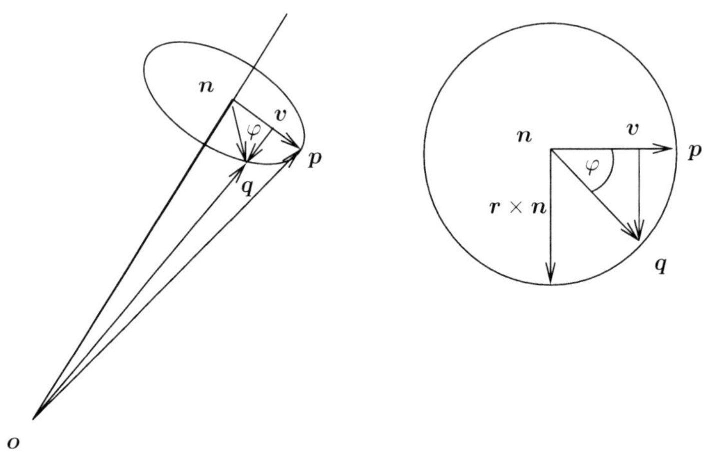

Fig. A.1 Axis angle representation of a rotation

The matrix $\boldsymbol{R}_{\boldsymbol{n},\theta}$ is

$$\boldsymbol{R}_{\boldsymbol{n},\theta} = \begin{pmatrix} (1-\cos\theta)n_x^2 + \cos\theta & (1-\cos\theta)n_x n_y + n_z \sin\theta & (1-\cos\theta)n_x n_z - n_y \sin\theta \\ (1-\cos\theta)n_x n_y - n_z \sin\theta & (1-\cos\theta)n_y^2 + \cos\theta & (1-\cos\theta)n_y n_z + n_x \sin\theta \\ (1-\cos\theta)n_x n_z + n_y \sin\theta & (1-\cos\theta)n_y n_z - n_x \sin\theta & (1-\cos\theta)n_z^2 + \cos\theta \end{pmatrix}.$$

A.1.3 Unit Quaternions

Quaternions are 4 dimensional complex numbers that may be used for representing rotations in 3D. They are generated by postulating an additional root of -1 that is named j and which is not i or $-i$. Then there exists necessarily the element k, such that $ij = k$, which is also a root of -1. The following relations hold true:

$$\begin{aligned} i^2 = j^2 &= k^2 = ijk &= -1 \\ ij = k, & \qquad\quad ji &= -k \\ jk = i, & \qquad\quad kj &= -i \\ ki = j, & \qquad\quad ik &= -j. \end{aligned}$$

i, j, k and -1 form a basis of the quaternion $\mathbb{H}$, that is named after the scientist Sir William Rowan Hamilton who had the idea on October 16 1843, while he was out walking along the Royal Canal in Dublin with his wife when the solution in the form of the equation $i^2 = j^2 = k^2 = ijk = -1$ suddenly occurred to him [55]. It is said, Hamilton then promptly carved this equation into the side of the nearby Broom Bridge.

All quaternion $\dot{\boldsymbol{q}}$ may be represented as $\dot{\boldsymbol{q}} = q_0 + q_x i + q_y j + q_z k$. Thus every quaternion defines uniquely a 4-vector $(q_0, q_x, q_y, q_z)^T \in \mathbb{R}^4$. A quaternion may be written as:

$$\dot{\boldsymbol{q}} = q_0 + q_x i + q_y j + q_z k = (q_0, q_x, q_y, q_z)^T = (q_0, \boldsymbol{q})^T = \begin{pmatrix} q_0 \\ q_y \\ q_x \\ q_z \end{pmatrix},$$

where real numbers are used as entries.

Similarly as the complex numbers, quaternions have the conjugation of the elements. The conjugate quaternion of $\dot{\boldsymbol{q}} = q_0 + q_x i + q_y j + q_z k$ is written as $\dot{\boldsymbol{q}}^*$ and is computed by negating all non-reel parts, i.e., $\dot{\boldsymbol{q}}^* = q_0 - q_x i - q_y j - q_z k = [q_0, -\boldsymbol{q}]$. Using the conjugate quaternion the norm of the quaternion is computed.

$$\|\dot{\boldsymbol{q}}\| = \sqrt{\langle \dot{\boldsymbol{q}}, \dot{\boldsymbol{q}}^* \rangle} = \sqrt{q_0^2 + q_x^2 + q_y^2 + q_z^2}.$$

Here the scalar product has been used. For quaternions, there are many kinds of multiplications defined, which is not in the scope of this section. We will need only the following non-commutative multiplication rule for two given quaternions $\dot{\boldsymbol{p}} = p_0 + p_x i + p_y j + p_z k$ and $\dot{\boldsymbol{q}} = q_0 + q_x i + q_y j + q_z k$:

$$\begin{aligned}
\dot{\boldsymbol{p}}\dot{\boldsymbol{q}} &= (p_0 + p_x i + p_y j + p_z k)(q_0 + q_x i + q_y j + q_z k) \\
&= (p_0 q_0 - p_x q_x - p_y q_y - p_z q_z) \\
&\quad + (p_0 q_x + p_x q_0 + p_y q_z - p_z q_y) i \\
&\quad + (p_0 q_y + p_x q_z + p_y q_0 - p_z q_x) j \\
&\quad + (p_0 q_z + p_x q_y + p_y q_x - p_z q_0) k \\
&= (p_0, \boldsymbol{p})(q_0, \boldsymbol{q}) \\
&= (p_0 q_0 - \boldsymbol{p} \cdot \boldsymbol{q}, \boldsymbol{p} \times \boldsymbol{q} + p_0 \boldsymbol{q} + q_0 \boldsymbol{p}).
\end{aligned}$$

A subset of the quaternions, namely the unit quternions are important, since they represent rotations. For a unit quaternion $\dot{\boldsymbol{q}}$o holds: $\|\dot{\boldsymbol{q}}\| = 1$. The unit quaternion has the property, that the inverse element $\dot{\boldsymbol{q}}^{-1}$ equals the conjugate elements $\dot{\boldsymbol{q}}^*$ using the above defined product. It is

$$\langle \dot{\boldsymbol{q}}, \dot{\boldsymbol{q}}^* \rangle = \|\dot{\boldsymbol{q}}\|^2 = 1.$$

Rotations using unit quaternions are represented as follows: Let $\dot{\boldsymbol{q}}$ be $\dot{\boldsymbol{q}} = (q_0, q_x, q_y, q_z) = (q_0, \boldsymbol{q})$. For the rotation of the 3D point $\boldsymbol{p} = (p_x, p_y, p_z)^T \in \mathbb{R}^3$ we write this point as quaternion, too, i.e., $\dot{\boldsymbol{p}} = (0, p_x, p_y, p_z)^T = (0, \boldsymbol{p})$. Then for the rotated point p_{rot} we have

$$\boldsymbol{p}_{\text{rot}} = \dot{\boldsymbol{q}}\dot{\boldsymbol{p}}\dot{\boldsymbol{q}}^*.$$

To show that the above formulation corresponds to a rotation, we rewrite the term to (For a detailed derivation please see [22].)

$$\begin{aligned}
p_{\text{rot}} &= \dot{\boldsymbol{q}}\dot{\boldsymbol{p}}\dot{\boldsymbol{q}}^* \\
&= (q_0, \boldsymbol{q})(0, \boldsymbol{p})(q_0, \boldsymbol{q})^* \\
&= (q_0, \boldsymbol{q})(0, \boldsymbol{p})(q_0, -\boldsymbol{q}) \\
&= (q_0, \boldsymbol{q})(\boldsymbol{q}, q_0 \boldsymbol{p} - \boldsymbol{p} \times \boldsymbol{q}) \\
&= (q_0(\boldsymbol{q} \cdot \boldsymbol{p}) - \boldsymbol{q} \cdot (q_0 \boldsymbol{p} - \boldsymbol{p} \times \boldsymbol{q}), \\
&\qquad q_0(q_0 \boldsymbol{p} - \boldsymbol{p} \times \boldsymbol{q}) + (\boldsymbol{q} \cdot \boldsymbol{p})\boldsymbol{q} + \boldsymbol{q} \times (q_0 \boldsymbol{p} - \boldsymbol{p} \times \boldsymbol{q})) \\
&= (0, q_0^2 \boldsymbol{p} + (\boldsymbol{q} \cdot \boldsymbol{p})\boldsymbol{q} - (\boldsymbol{q} \cdot \boldsymbol{q})\boldsymbol{p} + (\boldsymbol{q} \cdot \boldsymbol{p})\boldsymbol{q} + 2q_0(\boldsymbol{q} \times \boldsymbol{p})) \\
&= (0, (q_0^2 - \boldsymbol{q} \cdot \boldsymbol{q})\boldsymbol{p} + 2(\boldsymbol{q} \cdot \boldsymbol{p})\boldsymbol{q} + 2q_0(\boldsymbol{q} \times \boldsymbol{p})).
\end{aligned}$$

Rewriting the unit quaterion $\dot{\boldsymbol{q}}$ as $\dot{\boldsymbol{q}} = (\cos\theta, \sin\theta \boldsymbol{n})$ yields:

$$
\begin{aligned}
\boldsymbol{p}_{\text{rot}} &= (0, (\cos^2\theta - \sin^2\theta(\boldsymbol{n}\cdot\boldsymbol{n}))\boldsymbol{p} + 2((\sin\theta)\boldsymbol{n}\cdot\boldsymbol{p}) \\
&\qquad (\sin\theta)\boldsymbol{n} + 2\cos\theta((\sin\theta)\boldsymbol{n}\times\boldsymbol{p})) \\
&= (0, \cos 2\theta + (1-\cos 2\theta)(\boldsymbol{n}\cdot\boldsymbol{p})\boldsymbol{n} + (\boldsymbol{n}\times\boldsymbol{p})\sin 2\theta).
\end{aligned}
$$

The equation above has the same form as Eq. (A.1). Therefore unit quaternions are similar to the axis angle representation. To represent a rotation about the axis $\boldsymbol{n}$ with the angle θ using a unit quaternion $\dot{\boldsymbol{q}}$ one has to calculate:

$$
\dot{\boldsymbol{q}} = \left(\cos\frac{\theta}{2}, \sin\frac{\theta}{2}\boldsymbol{n}\right).
$$

To compute the rotation matrix $\boldsymbol{R}_q$ for a given quaternion $\dot{\boldsymbol{q}} = (q_0, q_x, q_y, q_z)$ we look at the components of $\boldsymbol{p}_{\text{rot}}$:

$$
\begin{aligned}
\boldsymbol{p}_{\text{rot}} &= \dot{\boldsymbol{q}}\dot{\boldsymbol{p}}\dot{\boldsymbol{q}}^* \\
&= \begin{pmatrix} 0 \\ q_0^2p_x + q_x^2p_x - (q_y^2+q_z^2)p_x + q_0(-2q_zp_y + 2q_yp_z) + 2q_x(q_yp_y + q_zp_z) \\ 2q_xq_yp_x + 2q_0q_zp_x + q_0^2p_y - q_x^2p_y + q_y^2p_y - q_z^2p_y - 2q_0q_xp_z + 2q_yq_zp_z \\ -2q_0q_yp_x + 2q_xq_zp_x + 2q_0q_xp_y + 2q_yq_zp_y + q_0^2p_z - q_x^2p_z - q_y^2p_z + q_z^2p_z \end{pmatrix} \\
&= \begin{pmatrix} 0 \\ p_{x,\text{rot}} \\ p_{y,\text{rot}} \\ p_{z,\text{rot}} \end{pmatrix}.
\end{aligned}
$$

This can be rewritten using matrix notation with the rotation matrix $\boldsymbol{R}_q$ as:

$$
\begin{aligned}
\boldsymbol{p}_{\text{rot}} &= \boldsymbol{R}_q\boldsymbol{p} \\
&= \begin{pmatrix} p_0^2 + p_x^2 - p_y^2 - p_z^2 & 2p_xp_y - 2p_0p_z & 2p_xp_z + 2p_0p_y \\ 2p_0p_z + 2p_xp_y & p_0^2 - p_x^2 + p_y^2 - p_z^2 & 2p_yp_z - 2p_0p_x \\ 2p_xp_z - 2p_0p_y & 2p_0p_x + 2p_yp_z & p_0^2 - p_x^2 - p_y^2 + p_z^2 \end{pmatrix} \begin{pmatrix} p_x \\ p_y \\ p_z \end{pmatrix} \\
&= \begin{pmatrix} p_{x,\text{rot}} \\ p_{y,\text{rot}} \\ p_{z,\text{rot}} \end{pmatrix}.
\end{aligned}
$$

A.1.4 Converting Rotation Representations

In many cases we have to convert one representation of a rotation into a different one, to exploit the advantages as appropriate. As already mentioned converting the axis angle representation of a rotation is performed by (rotation axis $\boldsymbol{n} = (n_x, n_y, n_z)^T$, angle θ):

$$\dot{\boldsymbol{q}} = \left(\cos\frac{\theta}{2}, \sin\frac{\theta}{2}n_x, \sin\frac{\theta}{2}n_y, \sin\frac{\theta}{2}n_z\right)^T = \left(\cos\frac{\theta}{2}, \sin\frac{\theta}{2}\boldsymbol{n}\right).$$

Converting a quaternion to axis angle representation is trival:

$$\begin{aligned} \theta &= 2\arccos q_0 \\ n_x &= \frac{q_x}{\sin\theta} \\ n_y &= \frac{q_y}{\sin\theta} \\ n_z &= \frac{q_z}{\sin\theta}. \end{aligned}$$

To derive a quaternion $\dot{\boldsymbol{q}} = (q_0, q_x, q_y, q_z)$ from the three Euler angles $(\theta_x, \theta_y, \theta_z)^T$ we use the quaternion multiplication rule. Firstly we represent the single rotations of the Euler angles by quaternions and compute the resulting quaternion b multiplication, afterwards. Since the Euler angle rotate around the axis of the coordinate system the quaternions $\dot{\boldsymbol{q}}_x$,$\dot{\boldsymbol{q}}_y$ und $\dot{\boldsymbol{q}}_z$ are:

$$\dot{\boldsymbol{q}}_x = \begin{pmatrix} \cos\frac{\theta_x}{2} \\ \sin\frac{\theta_x}{2} \\ 0 \\ 0 \end{pmatrix}, \dot{\boldsymbol{q}}_y = \begin{pmatrix} \cos\frac{\theta_y}{2} \\ 0 \\ \sin\frac{\theta_y}{2} \\ 0 \end{pmatrix}, \dot{\boldsymbol{q}}_z = \begin{pmatrix} \cos\frac{\theta_z}{2} \\ 0 \\ 0 \\ \sin\frac{\theta_z}{2} \end{pmatrix}.$$

The over-all rotation is given by the quaternion $\dot{\boldsymbol{q}}$ as:

$$\dot{\boldsymbol{q}} = \dot{\boldsymbol{q}}_z\dot{\boldsymbol{q}}_y\dot{\boldsymbol{q}}_x = \begin{pmatrix} \cos\frac{\theta_x}{2}\cos\frac{\theta_y}{2}\cos\frac{\theta_z}{2} + \sin\frac{\theta_x}{2}\sin\frac{\theta_y}{2}\sin\frac{\theta_z}{2} \\ \sin\frac{\theta_x}{2}\cos\frac{\theta_y}{2}\cos\frac{\theta_z}{2} - \cos\frac{\theta_x}{2}\sin\frac{\theta_y}{2}\sin\frac{\theta_z}{2} \\ \cos\frac{\theta_x}{2}\sin\frac{\theta_y}{2}\cos\frac{\theta_z}{2} + \sin\frac{\theta_x}{2}\cos\frac{\theta_y}{2}\sin\frac{\theta_z}{2} \\ \cos\frac{\theta_x}{2}\cos\frac{\theta_y}{2}\sin\frac{\theta_z}{2} - \sin\frac{\theta_x}{2}\sin\frac{\theta_y}{2}\cos\frac{\theta_z}{2} \end{pmatrix}.$$

The Euler angles corresponding to a quaternion are accordingly:

$$\begin{pmatrix} \theta_x \\ \theta_y \\ \theta_z \end{pmatrix} = \begin{pmatrix} \arctan\frac{2(p_0p_x+p_yp_z)}{1-2(p_x^2+p_y^2)} \\ \arcsin 2(p_0c - p_xp_z) \\ \arctan\frac{2(p_0p_z+p_xp_y)}{1-2(p_y^2+p_z^2)} \end{pmatrix}.$$

A.2 Plücker Coordinates

Given that G is a straight line in $\mathbb{R}^3$ and $\boldsymbol{p}$ an arbitrary point on G. Furthermore $\boldsymbol{g}$ is a direction vector and $\bar{\boldsymbol{g}} = \boldsymbol{p} \times \boldsymbol{g}$ is the momentum vector of the line G. The six entries of the concatenated vector $(\boldsymbol{g}, \bar{\boldsymbol{g}})$ are called Plücker coordinates of G. They have the following properties [59]:

- The momentum vector is independent of the choice of $\boldsymbol{p}$ on G. If one chooses an additional point $P\boldsymbol{p}' = \boldsymbol{p} + \lambda\boldsymbol{g}$ on the line G then it follows:

$$\boldsymbol{p}' \times \boldsymbol{g} = (\boldsymbol{p} + \lambda\boldsymbol{g}) \times \boldsymbol{g} = \boldsymbol{p} \times \boldsymbol{g}$$

- The Plücker Coordinates are homogeneous coordinates, since a direction vector $\boldsymbol{g}' = \lambda\boldsymbol{g}$ implies a momentum vector $\bar{\boldsymbol{g}}' = \boldsymbol{p} \times \boldsymbol{g}' = \lambda(\boldsymbol{p} \times \boldsymbol{g}) = \lambda\bar{\boldsymbol{g}}$. Therefore $\lambda(\boldsymbol{g}, \bar{\boldsymbol{g}})$ describes the same lines als $(\boldsymbol{g}, \bar{\boldsymbol{g}})$.
- The Plücker condition $\boldsymbol{g}^T\bar{\boldsymbol{g}} = 0$ holds true.

A.3 Additional Theorems from Linear Algebra

The following theorems originate from [62] (see also [32]) and generalize the theorems of section 4.1.1.

Lemma 7. *Given the orthonormal matrix $\boldsymbol{R}$. Then for every vector $\boldsymbol{x}$ holds:*

$$(\boldsymbol{Rx}) \cdot \boldsymbol{x} \leq \boldsymbol{x} \cdot \boldsymbol{x}.$$

The two terms are equal, if $\boldsymbol{Rx} = \boldsymbol{x}$.

Proof. It holds

$$(\boldsymbol{Rx}) \cdot (\boldsymbol{Rx}) = (\boldsymbol{Rx})^T(\boldsymbol{Rx}) = \boldsymbol{x}^T\boldsymbol{R}^T\boldsymbol{Rx} = \boldsymbol{x}^T\boldsymbol{x} = \boldsymbol{x} \cdot \boldsymbol{x},$$

since $\boldsymbol{R}^T\boldsymbol{R} = \mathbb{1}$. Furthermore we have

$$\begin{aligned}(\boldsymbol{Rx} - \boldsymbol{x}) \cdot (\boldsymbol{Rx} - \boldsymbol{x}) &= (\boldsymbol{Rx}) \cdot (\boldsymbol{Rx}) - 2(\boldsymbol{Rx}) \cdot \boldsymbol{x} + \boldsymbol{x} \cdot \boldsymbol{x} \\ &= 2(\boldsymbol{x} \cdot \boldsymbol{x} - (\boldsymbol{Rx}) \cdot \boldsymbol{x}).\end{aligned} \tag{A.2}$$

Because $(\boldsymbol{Rx} - \boldsymbol{x}) \cdot (\boldsymbol{Rx} - \boldsymbol{x}) \geq 0$ the lemma is proven. The equality if given, if $(\boldsymbol{Rx} - \boldsymbol{x}) \cdot (\boldsymbol{Rx} - \boldsymbol{x}) = 0$ is the case. According to Eq. (A.2) this only happens if, $\boldsymbol{Rx} = \boldsymbol{x}$. □

Lemma 8. *Every positive semi-definite $n \times n$ matrix $\boldsymbol{S}$ can be written using the orthonormal set of vectors $\{\boldsymbol{u}_i\}$ as*

$$\boldsymbol{S} = \sum_{i=1}^{n} \boldsymbol{u}_i\boldsymbol{u}_i^T.$$

Proof. If the eigenvalues of $\boldsymbol{S}$ are $\{\lambda_i\}$. The corresponding set of eigenvectors are $\{\hat{\boldsymbol{u}}_i\}$. Then matrix $\boldsymbol{S}$ is given by

$$\boldsymbol{S} = \sum_{i=1}^{n} \lambda_i \hat{\mathbf{u}}_i \hat{\mathbf{u}}_i^T .$$

Since, the eigenvalues of a positive semi-definite matrix are not negative, the claim of the lemma is derived with $\boldsymbol{u}_i = \sqrt{\lambda_i}\hat{\mathbf{u}}_i$. □

Lemma 9. *Suppose $\boldsymbol{S}$ is a positive semi-definite matrix. The for all orthonormal matrices $\boldsymbol{R}$ holds*

$$tr(\boldsymbol{RS}) \leq tr(\boldsymbol{S}) ,$$

where the equality is the case, if $\boldsymbol{RS} = \boldsymbol{S}$.

Proof. If $\boldsymbol{S}$ is represented according to lemma 8. Since $\text{tr}\left(\boldsymbol{a}\boldsymbol{b}^T\right) = \boldsymbol{a}\cdot\boldsymbol{b}$ holds, it follows

$$\text{tr}\left(\boldsymbol{S}\right) = \sum_{i=1}^{n} \boldsymbol{u}_i \cdot \boldsymbol{u}_i \qquad \text{and} \qquad \text{tr}\left(\boldsymbol{RS}\right) = \sum_{i=1}^{n} (\boldsymbol{R}\boldsymbol{u}_i) \cdot \boldsymbol{u}_i ,$$

where the last equation is according to lemma 7 smaller or equal of the trace $\text{tr}\left(\boldsymbol{S}\right)$. Equality if given for $\boldsymbol{R}\boldsymbol{u}_i = \boldsymbol{u}_i$, i.e., if $\boldsymbol{RS} = \boldsymbol{S}$, because $\boldsymbol{S}\boldsymbol{u}_i = \boldsymbol{u}_i$ holds. □

Corollary 1. *If $\boldsymbol{S}$ is a positive definite matrix, then it holds for every orthonormal matrix $\boldsymbol{R}$*

$$tr(\boldsymbol{RS}) \leq tr(\boldsymbol{S}) .$$

The equality is the case, if $\boldsymbol{R} = \mathbb{1}$, i.e, $\boldsymbol{R}$ is the identity matrix. □

Lemma 10. *The matrix $\boldsymbol{T}$ is the positive semi-definite square root of the positive definite matrix $\boldsymbol{S}$:*

$$\boldsymbol{T} = \sum_{i=1}^{n} \sqrt{\lambda_i} \hat{\mathbf{u}}_i \hat{\mathbf{u}}_i^T \qquad \textit{and} \qquad \boldsymbol{S} = \sum_{i=1}^{n} \lambda_i \hat{\mathbf{u}}_i \hat{\mathbf{u}}_i^T ,$$

where $\{\lambda_i\}$ is the set of eigenvalues and $\{\hat{\mathbf{u}}_i\}$ the set of orthonormal unit eigenvectors of $\boldsymbol{S}$.

Proof. We have

$$\boldsymbol{T}^2 = \left(\sum_{i=1}^{n} \sqrt{\lambda_i} \hat{\mathbf{u}}_i \hat{\mathbf{u}}_i^T\right) \left(\sum_{j=1}^{n} \sqrt{\lambda_j} \hat{\mathbf{u}}_j \hat{\mathbf{u}}_j^T\right) = \sum_{i=1}^{n} \sum_{j=1}^{n} \sqrt{\lambda_i \lambda_j} \left(\hat{\mathbf{u}}_i \cdot \hat{\mathbf{u}}_j\right) \hat{\mathbf{u}}_i \hat{\mathbf{u}}_j^T$$

$$= \sum_{k=1}^{n} \lambda_k \hat{\mathbf{u}}_k \hat{\mathbf{u}}_k^T = \boldsymbol{S},$$

since $\hat{\boldsymbol{u}}_i \cdot \hat{\boldsymbol{u}}_j = 0$ holds for $i \neq j$. Furthermore we have

$$\boldsymbol{x}\boldsymbol{T}\boldsymbol{x} = \sum_{i=1}^{n} \lambda_i (\hat{\boldsymbol{u}}_i \cdot \boldsymbol{x})^2 \leq 0,$$

since the eigenvalues are $\lambda_i \geq 0$. Therefore $\boldsymbol{T}$ positive semi-definite. □

Corollary 2. *There exists a positive semi-definite square root matrix for all positive semi-definite matices.* □

Corollary 3. *The matrix*

$$\boldsymbol{T}^{-1} = \sum_{i=1}^{n} \frac{1}{\sqrt{\lambda_i}} \hat{\boldsymbol{u}}_i \hat{\boldsymbol{u}}_i^T$$

is the inverse of the positive semi-definite square root matrix of the positive semi-definite matrix

$$\boldsymbol{S} = \sum_{i=1}^{n} \lambda_i \hat{\boldsymbol{u}}_i \hat{\boldsymbol{u}}_i^T.$$

□

Lemma 11. *For all matrices $\boldsymbol{X}$ the following matrices are positive semi-definite*

$$\boldsymbol{X}^T\boldsymbol{X} \qquad \text{and} \qquad \boldsymbol{X}\boldsymbol{X}^T.$$

Proof. The matrix $\boldsymbol{X}^T\boldsymbol{X}$ is symmetric, since $(\boldsymbol{X}^T\boldsymbol{X})^T = \boldsymbol{X}^T(\boldsymbol{X}^T)^T = \boldsymbol{X}^T\boldsymbol{X}$ holds. Furthermore for all vectors $\boldsymbol{x}$ we have

$$\boldsymbol{x}^T(\boldsymbol{X}^T\boldsymbol{X})\boldsymbol{x} = (\boldsymbol{x}^T\boldsymbol{X}^T)(\boldsymbol{X}\boldsymbol{x}) = (\boldsymbol{X}\boldsymbol{x})^T(\boldsymbol{X}\boldsymbol{x}) = (\boldsymbol{X}\boldsymbol{x}) \cdot (\boldsymbol{X}\boldsymbol{x}) \geq 0.$$

An analogous argumentation holds for $\boldsymbol{X}\boldsymbol{X}^T$. □

Corollary 4. *For all non-singular square matrix $\boldsymbol{X}$ the following matrices are positive definite:*

$$\boldsymbol{X}^T\boldsymbol{X} \qquad \text{and} \qquad \boldsymbol{X}\boldsymbol{X}^T.$$

□

Lemma 12. *Every non-singular matrix $\boldsymbol{X}$ may be written as*

$$\boldsymbol{X} = \boldsymbol{P}\boldsymbol{S}$$

There we have

$$\boldsymbol{P} = \boldsymbol{X}(\boldsymbol{X}^T\boldsymbol{X})^{-1/2}$$

where $\boldsymbol{P}$ is an orthonormal matrix and

$$\boldsymbol{S} = (\boldsymbol{X}^T\boldsymbol{X})^{-1/2}$$

a positive semi-definite matrix.

Proof. Since $\boldsymbol{X}$ is non-singular we know according to corollary 4 that $\boldsymbol{X}^T\boldsymbol{X}$ is positive definite. In addition we have with corollary 2 the existence of the positive definite square root matrix $(\boldsymbol{X}^T\boldsymbol{X})^{1/2}$, that is constructed by corollary 3. Thus we can compute $\boldsymbol{S}$ and $\boldsymbol{P}$ for given matrix $\boldsymbol{X}$. Obviously their product is

$$\boldsymbol{P}\boldsymbol{S} = \boldsymbol{X}(\boldsymbol{X}^T\boldsymbol{X})^{-1/2}(\boldsymbol{X}^T\boldsymbol{X})^{-1/2} = \boldsymbol{X}.$$

We have to check, if $\boldsymbol{P}$ is orthonormal. It holds:

$$\boldsymbol{P}^T = (\boldsymbol{X}^T\boldsymbol{X})^{-1/2}\boldsymbol{X}^T$$

and therefore

$$\begin{aligned}\boldsymbol{P}^T\boldsymbol{P} &= (\boldsymbol{X}^T\boldsymbol{X})^{-1/2}(\boldsymbol{X}^T\boldsymbol{X})(\boldsymbol{X}^T\boldsymbol{X})^{-1/2}\\ &= (\boldsymbol{X}^T\boldsymbol{X})^{-1/2}(\boldsymbol{X}^T\boldsymbol{X})^{1/2}(\boldsymbol{X}^T\boldsymbol{X})^{1/2}(\boldsymbol{X}^T\boldsymbol{X})^{-1/2}\\ &= \mathbb{1}.\end{aligned}$$ □

A.4 Solving Linear Systems of Equations

Assume we have to solve the following system of linear equations

$$\boldsymbol{A}\hat{\boldsymbol{x}} \approx \boldsymbol{b}, \tag{A.3}$$

where $\boldsymbol{A}$ is a non-square matrix, for instance the system might be over-determined. Therefore inverting $\boldsymbol{A}$ is not possible and one has to find the optimal solution to the following minimization problem:

$$\|\boldsymbol{A}\boldsymbol{x} - \boldsymbol{b}\|^2 = 0 \tag{A.4}$$

Eq. (A.4) is the linear least square formulation of the solution to the linear equation (A.3). To solve (A.4) we firstly remove the square term by

$$\|\boldsymbol{A}\boldsymbol{x} - \boldsymbol{b}\|^2 = ((\boldsymbol{A}\boldsymbol{x})_1 - b_1)^2 + ((\boldsymbol{A}\boldsymbol{x})_2 - b_2)^2 + \cdots((\boldsymbol{A}\boldsymbol{x})_n - b_n)^2$$

where the subscript denoted the entry of the vector. Thus

$$\begin{aligned}&= (\boldsymbol{A}\boldsymbol{x} - \boldsymbol{b})^T(\boldsymbol{A}\boldsymbol{x} - \boldsymbol{b})\\ &= (\boldsymbol{A}\boldsymbol{x})^T(\boldsymbol{A}\boldsymbol{x}) - 2(\boldsymbol{A}\boldsymbol{x})^T\boldsymbol{b} + \boldsymbol{b}^T\boldsymbol{b}\end{aligned}$$

To find the minimum we take the derivative and set it to zero. Thus we have

$$\frac{d}{d\boldsymbol{x}}(\boldsymbol{A}\boldsymbol{x})^T(\boldsymbol{A}\boldsymbol{x}) - 2(\boldsymbol{A}\boldsymbol{x})^T\boldsymbol{b} = 2\boldsymbol{A}^T\boldsymbol{A}\hat{\boldsymbol{x}} - 2\boldsymbol{A}\boldsymbol{b} = \boldsymbol{0}.$$

and

$$A^T A\hat{x} = A^T b.$$

The solution is given by the so-called pseudo-inverse as

$$\hat{x} = (A^T A)^{-1} A^T b. \tag{A.5}$$

Since $(A^T A)$ is a square matrix we have good chance to be able to carry out the matrix inversion. There are many algorithms available for solving (A.5), e.g.,

Choleskey Decomposition. If A is symmetric and positive definite, then we can solve $Ax = b$ by first computing the Cholesky decomposition $A = LL^T$, then solving $Ly = b$ for y, and finally solving $L^T x = y$ for x.

QR Decomposition. Of a real square matrix A is a decomposition of A as $A = QR$, where Q is an orthogonal matrix and R is an upper triangular matrix. Similarly to the Choleskey decomposition the solution of (A.3) is found. Several strategies exist to compute the QR decomposition, namely by the Gram-Schmidt method, by means of Householder reflections, and by means of Givens rotations.

Singular Value Decomposition. (SVD) is the most stable method to compute the solution to (A.3). Here $x = V\Sigma^+ U^* b$, where V, Σ, and U are the output of the SVD as follows. If the given matrix A is a $m \times n$ matrix, then there exists a factorization of the form

$$M = U\Sigma V^*,$$

where U is a $m \times m$ unitary matrix. The matrix Σ has size $m \times n$ and contains nonnegative numbers on the diagonal in non-increasing fashion and zeros off the diagonal. V^* denotes the $n \times n$ conjugate transpose of V, The matrices V, Σ, and U have the following properties:

- The matrix V contains a set of orthonormal "input" or "analysing" basis vector directions for A.
- The matrix U contains a set of orthonormal "output" basis vector directions for A.
- The matrix Σ contains the singular values, that can be thought of as scalar "gain controls" by which each corresponding input is multiplied to give a corresponding output.

A.5 Computing Constrained Extrema

Optimizing with subject to conditions means finding the maximum or minimum of f for given functions $f : \chi \to \mathbb{R}$ and $g : \chi \to \mathbb{R}^q$ and a given vector $c \in \mathbb{R}^q$ such that the conditions $g = c$ or $g \leq c$ by components hold. In most cases it is easier to minimize the function $f + \lambda^T g$ over the whole space χ and vary the constant λ afterwards. The following theorem shows why and how this works [36].

Theorem 1. *Assume for a given* $\boldsymbol{\lambda} \in \mathbb{R}^q$ $\boldsymbol{x}_{\boldsymbol{\lambda}} \in \arg\min_{\boldsymbol{x}\in\chi_{\boldsymbol{\lambda}}} \left(f(\boldsymbol{x})+\boldsymbol{\lambda}^T\boldsymbol{g}(\boldsymbol{x})\right)$. *With* $\boldsymbol{c\lambda} = \boldsymbol{g}(\boldsymbol{x}_{\boldsymbol{\lambda}}))$ *we have*

$$\boldsymbol{x}_{\boldsymbol{\lambda}} \in \arg \min_{\boldsymbol{x}\in\chi:g(\boldsymbol{x})=c_{\boldsymbol{\lambda}}} f(\boldsymbol{x}).$$

Now assume $\chi_{\boldsymbol{\lambda}}$ *to be an arbitrary subset of* χ, *such that* $\boldsymbol{x}_{\boldsymbol{\lambda}} \in \chi_{\boldsymbol{\lambda}}$ *and*

$$g_i(\boldsymbol{x}) \begin{Bmatrix} \leq c_{\lambda_i} & , \textit{if} \quad \lambda_i > 0 \\ \geq c_{\lambda_i} & , \textit{if} \quad \lambda_i < 0 \end{Bmatrix} \quad \textit{for all } \boldsymbol{x} \in \chi_{\boldsymbol{\lambda}}.$$

Then it follows

$$\boldsymbol{x}_{\boldsymbol{\lambda}} \in \arg\min_{\boldsymbol{x}\in\chi_{\lambda}} f(\boldsymbol{x}).$$

If $\boldsymbol{x}_{\boldsymbol{\lambda}}$ *is the unique minimum of* $f + \boldsymbol{\lambda}^T\boldsymbol{g}$ *over* χ, *then* $\boldsymbol{x}_{\boldsymbol{\lambda}}$ *is also the unique minimum of* f *over* $\chi_{\boldsymbol{\lambda}}$.

For the given problem of minimizing f subject to $\boldsymbol{g} = \boldsymbol{c}$ we have the following procedure: One minimizes $f + \boldsymbol{\lambda}^T\boldsymbol{g}$ over χ. Assume $\boldsymbol{x}_\lambda$ is such a minimum. One has to choose $\boldsymbol{\lambda}$ such that $\boldsymbol{g}(x_{\boldsymbol{\lambda}}) = \boldsymbol{c}$. In this case $\boldsymbol{x}_{\boldsymbol{\lambda}}$ is the solution of the given problem. Replacing the side condition $\boldsymbol{g} = \boldsymbol{c}$ with the component wise inequation $\boldsymbol{g} \leq \boldsymbol{c}$, then we have to choose $\boldsymbol{\lambda}$ such that $\boldsymbol{g}(\boldsymbol{x}_{\boldsymbol{\lambda}}) = \boldsymbol{c}$ and $\boldsymbol{\lambda} \geq 0$.

Proof. Let $\boldsymbol{y}$ be an arbitrary point in $\chi_{\boldsymbol{\lambda}}$. Suppose $f(\boldsymbol{y}) \leq f(\boldsymbol{x}_{\boldsymbol{\lambda}})$, then it is

$$\begin{aligned} f(\boldsymbol{y}) + \boldsymbol{\lambda}^T g(\boldsymbol{y}) &= f(\boldsymbol{y}) + \sum_{i=1}^{q} \underbrace{\lambda_i g_i(\boldsymbol{y})}_{\leq \lambda_i c_{\lambda_i}} \\ &\leq f(\boldsymbol{x}_\lambda) + \sum_{i=1}^{q} \lambda_i c_{\lambda_i} \\ &= f(\boldsymbol{x}_\lambda) + \lambda^T g(\boldsymbol{x}_\lambda). \end{aligned}$$

From the optimality of $\boldsymbol{x}_{\boldsymbol{\lambda}}$ it follows that $f(\boldsymbol{y}) = f(\boldsymbol{x}_{\boldsymbol{\lambda}})$. If $\boldsymbol{x}_{\boldsymbol{\lambda}}$ is the unique minimum of $f + \boldsymbol{\lambda}^T\boldsymbol{g}$ over χ, then from the inequality $f(\boldsymbol{y}) \leq f(\boldsymbol{x}_{\boldsymbol{\lambda}})$ it follows that $\boldsymbol{y} = \boldsymbol{x}_{\boldsymbol{\lambda}}$. □

A.5.1 Computing the Minimum of a Quadratic Form with Subject to Conditions

Assume $\boldsymbol{A} \in \mathbb{R}^{d\times d}$ to be symmetric and positive definite. The task is now to minimize

$$f(\boldsymbol{x}) = \frac{1}{2}\boldsymbol{x}^T\boldsymbol{A}\boldsymbol{x}$$

for all $\boldsymbol{x} \in \mathbb{R}^d$ subject to

$$\boldsymbol{B}\boldsymbol{x} = \boldsymbol{c}.$$

Here $\boldsymbol{B}$ is a matrix in $\mathbb{R}^{q \times d}$ with rank $q \leq d$ and $\boldsymbol{c}$ is a vector in $\mathbb{R}^q$.

For that purpose we minimize for a $\boldsymbol{\lambda} \in \mathbb{R}^d$ the function $f(\boldsymbol{x}) = \boldsymbol{\lambda}^T \boldsymbol{B}\boldsymbol{x}$. It holds:

$$\begin{aligned} f(\boldsymbol{x}) + \boldsymbol{\lambda}^T \boldsymbol{B}\boldsymbol{x} &= 2^{-1}\boldsymbol{x}^T \boldsymbol{A}\boldsymbol{x} + (\boldsymbol{B}^T\boldsymbol{\lambda})^T \boldsymbol{x} \\ &= 2^{-1}\left(\boldsymbol{x}^T \boldsymbol{A}\boldsymbol{x} + 2(\boldsymbol{A}^{-1}\boldsymbol{B}^T\boldsymbol{\lambda})^T \boldsymbol{A}\boldsymbol{x}\right) \\ &= 2^{-1}(\boldsymbol{x} + \boldsymbol{A}^{-1}\boldsymbol{B}^T\boldsymbol{\lambda})^T \boldsymbol{A}(\boldsymbol{x} + \boldsymbol{A}^{-1}\boldsymbol{B}^T\boldsymbol{\lambda}) - 2^{-1}\boldsymbol{\lambda}^T \boldsymbol{B}\boldsymbol{A}^{-1}\boldsymbol{B}^T\boldsymbol{\lambda}. \end{aligned}$$

This term is minimal if and only if

$$\boldsymbol{x}_{\boldsymbol{\lambda}} = -\boldsymbol{A}^{-1}\boldsymbol{B}^T\boldsymbol{\lambda}$$

holds. Furthermore it is

$$\boldsymbol{B}\boldsymbol{x}_{\lambda} = -\boldsymbol{B}\boldsymbol{A}^{-1}\boldsymbol{B}^T\lambda$$

and this is equal $\boldsymbol{c}$, iff

$$\lambda = -(\boldsymbol{B}\boldsymbol{A}^{-1}\boldsymbol{B}^T)^{-1}\boldsymbol{c}.$$

Accordingly the given problem has a unique solution at

$$\boldsymbol{x} = \boldsymbol{A}^{-1}\boldsymbol{B}^T(\boldsymbol{B}\boldsymbol{A}^{-1}\boldsymbol{B}^T)^{-1}\boldsymbol{c}$$

with

$$f(\boldsymbol{x}) = \frac{1}{2}\boldsymbol{c}^T(\boldsymbol{B}\boldsymbol{A}^{-1}\boldsymbol{B}^T)^{-1}\boldsymbol{c}.$$

It seams that the method of Lagrange is a trick that works sometimes, we have to check when and why it works. Using convex analysis we can show that the method has to be correct in many cases.

Theorem 2. *If χ is a convex and open subset of $\mathbb{R}^d$, assume $f : \chi \to \mathbb{R}$ is a convex and $\boldsymbol{g} : \mathbb{R}^d \to \mathbb{R}^q$ a linear function. Suppose for a given $\boldsymbol{c} \in \mathbb{R}^q$ exists a*

$$\boldsymbol{x}_{\boldsymbol{c}} \in \arg \min_{x \in \chi : \boldsymbol{g}(x) = \boldsymbol{c}} f(x).$$

Then there exists a vector $\boldsymbol{\lambda}_c \in \mathbb{R}^q$, such that

$$\boldsymbol{x}_{\boldsymbol{c}} \in \arg\max_{\boldsymbol{x} \in \chi} \left(f(x) + \boldsymbol{\lambda}_c^T g(\boldsymbol{x})\right) \quad \textit{holds.}$$

Proof. Let K be the set of all $\boldsymbol{x} \in \chi$, such that $g(\boldsymbol{x}) = \boldsymbol{c}$. According to the precondition

$$E(f) = \{(x, r) \in \chi \times \mathbb{R} : f(\boldsymbol{x}) \le r\} \quad \text{and} \quad D = K \times] - \infty, f(\boldsymbol{x_c})[$$

are non-empty convex subsets of $\mathbb{R}^d \times \mathbb{R}$. Now the theorem of Eidelheit gives the existence of a pair $(\boldsymbol{v}, t) \in \mathbb{R}^d \times \mathbb{R} \setminus \{(0, 0)\}$, such that for all $(\boldsymbol{x}, r) \in E(f)$, $(\boldsymbol{y}, s) \in D$ holds:

$$\boldsymbol{v}^T \boldsymbol{x} + t\, r \ge \boldsymbol{v}^T \boldsymbol{y} + ts.$$

The scalar t cannot be negative, since otherwise the right part of the inequality would be arbitrary large if $f(\boldsymbol{x_c}) \ge s \to \infty$. In addition the case $t = 0$ must not be considered, since otherwise it is $\boldsymbol{v} \neq \boldsymbol{0}$ and $\boldsymbol{v}^T \boldsymbol{y} \ge \boldsymbol{v}^T \boldsymbol{x_c}$ for all $\boldsymbol{x} \in \chi$, which contradicts the precondition that χ is open. Therefore $t > 0$ and without loss of generality we assume $t = 1$. Now it holds:

$$\boldsymbol{v}^T \boldsymbol{y} + f(\boldsymbol{x_c}) \ge \boldsymbol{v}^T \boldsymbol{y} + f(\boldsymbol{x_c}) \quad \text{for all } \boldsymbol{x} \in \chi, \boldsymbol{y} \in K. \tag{A.6}$$

Setting $\boldsymbol{x} = \boldsymbol{x_c}$ on the right side of (A.6) then we have

$$\boldsymbol{v}^T \boldsymbol{y} \le \boldsymbol{v}^T \boldsymbol{x}_c \tag{A.7}$$

for all $\boldsymbol{y} \in K$. For a sufficient small $\delta > 0$ all points where $\boldsymbol{y} = \boldsymbol{x}_c \pm w$ with $\boldsymbol{w} \in \mathbb{R}^d$, $g(\boldsymbol{w}) = 0$ and $||\boldsymbol{w}|| < \delta$ belong to the set K. Substituting these points in Eq. (A.7), it turns out that $\boldsymbol{v}^T \boldsymbol{w} = 0$ for all points $\boldsymbol{w}$ with $\boldsymbol{g}(\boldsymbol{w}) = 0$. Now let $\boldsymbol{g}(\boldsymbol{w}) = \boldsymbol{B}\boldsymbol{w}$ with a matrix $\boldsymbol{B} \in \mathbb{R}^{q \times d}$. Rewriting

$$\boldsymbol{B} = (\boldsymbol{b}_1, \boldsymbol{b}_2, \ldots, \boldsymbol{b}_q)^T$$

with vectors $\boldsymbol{b}_1, \ldots, \boldsymbol{b}_q \in \mathbb{R}^d$, we have that vector $\boldsymbol{v}$ is perpendicular to all vectors in $\{\boldsymbol{b}_1, \ldots, \boldsymbol{b}_q\}$. But this means that $\boldsymbol{v}$ is a linear combination of the vectors $\boldsymbol{b}_1, \ldots, \boldsymbol{b}_q$. In other words, $\boldsymbol{v} = \sum_{i=1}^{q} \lambda_i \boldsymbol{b}_i = \boldsymbol{B}^T \boldsymbol{\lambda}$ for a $\boldsymbol{\lambda} \in \mathbb{R}^q$. Therefore $\boldsymbol{v}^T \boldsymbol{x} = \boldsymbol{\lambda}^T g(\boldsymbol{x})$. From Eq. (A.6) we have in addition for $\boldsymbol{y} = \boldsymbol{x}_c$ the inequality

$$f(\boldsymbol{x}) + \boldsymbol{\lambda}^T \boldsymbol{g}(x) \ge f(\boldsymbol{x_c}) + \boldsymbol{\lambda}^T g(\boldsymbol{x_c}) \quad \text{for all } \boldsymbol{x} \in \chi. \qquad \square$$

A.6 Numerical Function Minimization

Powell's Method

Powell's method computes directions for function minimization in one direction [99]. From the starting point $\boldsymbol{p}_0$ in the n-dimensional search space (the concatenation of the 3-vector descriptions of all planes) the error function (7.2) is optimized along a direction $\boldsymbol{i}$ using a one dimensional minimization method, e.g., Brent's method [100].

Conjugate directions are good search directions, while unit basis directions are inefficient in error functions with valleys. At the line minimum of a function along the direction $\boldsymbol{i}$ the gradient is perpendicular to $\boldsymbol{i}$. In addition, the n-dimensional function is approximated at point $\boldsymbol{p}$ by a Taylor series using point $\boldsymbol{p}_0$ as origin of the coordinate system. It is

$$E(\boldsymbol{p}) = E(\boldsymbol{p}_0) + \sum_k \frac{\partial E}{\partial \boldsymbol{p}_k} \boldsymbol{p}_k + \sum_{k,l} \frac{\partial^2 E}{\partial \boldsymbol{p}_k \partial \boldsymbol{p}_l} \boldsymbol{p}_k \boldsymbol{p}_l + \cdots$$
$$\approx c - \boldsymbol{b} \cdot \boldsymbol{p} + \frac{1}{2} \boldsymbol{p} \cdot \boldsymbol{A} \cdot \boldsymbol{p}$$

with $c = E(\boldsymbol{p}_0)$, $\boldsymbol{b} = \nabla E|_{\boldsymbol{p}_0}$ and $\boldsymbol{A}$ the Hessian matrix of E at point $\boldsymbol{p}_0$. Given a direction $\boldsymbol{i}$, the method of conjugate gradients selects a new direction $\boldsymbol{j}$ so that $\boldsymbol{i}$ and $\boldsymbol{j}$ are perpendicular. This selection prevents interference of minimization directions. For the approximation above the gradient of E is $\nabla E = \boldsymbol{A} \cdot \boldsymbol{p} - \boldsymbol{b}$. From the differentiation ($\delta(\nabla E) = \boldsymbol{A}(\delta \boldsymbol{p})$) it follows for directions $\boldsymbol{i}$ and $\boldsymbol{j}$ that

$$0 = \boldsymbol{i} \cdot \delta(\nabla E) = \boldsymbol{i} \cdot \boldsymbol{A} \cdot \boldsymbol{j}.$$

With the above equation conjugate directions are defined and Powell's method produces such directions, without computing derivatives.

The following heuristic scheme is implemented for finding new directions. Starting point is the description of the planes and the initial directions $\boldsymbol{i}_l$, $l = 1, \ldots, n$ are the unit basis directions. The algorithm repeats the following steps until the error function (7.2) reaches a minimum [100]:

1. Save the starting position as $\boldsymbol{p}_0$.
2. For $l = 1, \ldots, n$, minimize the error function (7.2) starting from $\boldsymbol{p}_{l-1}$ along the direction $\boldsymbol{i}_l$ and store the minimum as the next position $\boldsymbol{p}_l$. After the loop, all $\boldsymbol{p}_l$ are computed.
3. Let $\boldsymbol{i}_l$ be the direction of the largest decrease. Now this direction $\boldsymbol{i}_l$ is replaced with the direction given by $(\boldsymbol{p}_n - \boldsymbol{p}_0)$. The assumption of the heuristic is that the substituted direction includes the replaced direction so that the resulting set of directions remains linear independent.
4. The iteration process continues with the new starting position $\boldsymbol{p}_0 = \boldsymbol{p}_n$, until the minimum is reached.

Downhill Simplex Method

Another suitable optimization algorithm for eq. (7.2) is the downhill simplex method as used by Cantzler et al. [15]. A nondegenerate simplex is a geometrical figure consisting of $N + 1$ vertices in N dimensions, whereas the $N + 1$ vertices span a N-dimensional vector space. Given an initial starting point $\boldsymbol{p}_0$, the starting simplex is computed through

$$\boldsymbol{p}_i = \boldsymbol{p}_0 + \lambda \boldsymbol{i}_l,$$

with $\boldsymbol{i}_l$ the unit basis directions and λ a constant that depends on the problem's characteristic length scale [100]. In our experiments λ is set to 0.15.

The downhill simplex method consists of a series of steps, i.e., reflections and contractions [100]. In a reflection step the algorithm moves the point of the simplex where the function is largest through the opposite face of the simplex to some lower point. If the algorithm reaches a "valley floor", the method contracts the simplex, i.e., the volume of the simplex decreases by moving one or several points, and moves along the valley [100].

References

1. The RANSAC (Random Sample Consensus) Algorithm (2003), http://www.dai.ed.ac.uk/CVonline/LOCAL_COPIES/FISHER/RANSAC/
2. Allen, P., Stamos, I., Gueorguiev, A., Gold, E., Blaer, P.: AVENUE: Automated Site Modelling in Urban Environments. In: Proceedings of the Third International Conference on 3D Digital Imaging and Modeling (3DIM 2001). Quebec City, Canada (2001)
3. Arun, K.S., Huang, T.S., Blostein, S.D.: Least square fitting of two 3-d point sets. IEEE Transactions on Pattern Analysis and Machine Intelligence 9(5), 698–700 (1987)
4. Arya, S., Mount, D.M., Netanyahu, N.S., Silverman, R., Wu, A.Y.: An Optimal Algorithms for Approximate Nearest Neighbor Searcching in Fixed Dimensions. Journal of the ACM 45, 891–923 (1998)
5. Bailey, T.: Mobile Robot Localisation and Mapping in Extensive Outdoor Environments. PhD thesis, University of Sydney, Sydney, NSW, Australia (2002)
6. Belwood, J.J., Waugh, R.J.: Bats and mines: Abandoned does not always mean empty. Bats 9(3) (1991)
7. Benjemaa, R., Schmitt, F.: Fast Global Registration of 3D Sampled Surfaces Using a Multi-Z-Buffer Technique. In: Proceedings IEEE International Conference on Recent Advances in 3D Digital Imaging and Modeling (3DIM 1997), Ottawa, Canada (May 1997)
8. Benjemaas, R., Schmitt, F.: A Solution For The Registration Of Multiple 3D Point Sets Using Unit Quaternions. In: Burkhardt, H., Neumann, B. (eds.) ECCV 1998. LNCS, vol. 1407, pp. 34–50. Springer, Heidelberg (1998)
9. Bentley, J.L.: Multidimensional binary search trees used for associative searchin. Communications of the ACM 18(9), 509–517 (1975)
10. Bergevin, R., Soucy, M., Gagnon, H., Laurendeau, D.: Towards a general multi-view registration technique. IEEE Transactions on Pattern Analysis and Machine Intelligence (PAMI) 18(5), 540–547 (1996)
11. Besl, P., McKay, N.: A method for Registration of 3–D Shapes. IEEE Transactions on Pattern Analysis and Machine Intelligence 14(2), 239–256 (1992)
12. Borenstein, J., Everett, H.R., Feng, L.: Where am I? Sensors and Methods for Mobile Robot Positioning. University of Michigan (April 1996)

13. Borrmann, D., Elseberg, J.: Global Konsistente 3D Kartierung am Beispiel des Botanischen Gartens in Osnabrück. Bachelor's thesis, Universität Osnabrück (October 2006)
14. Boulanger, P., Jokinen, O., Beraldin, A.: Intrinsic Filtering of Range Images Using a Physically Based Noise Model. In: Proceedings of the 15th International Conference on Vision Interface, Calgary, Canada, pp. 320–331 (May 2002)
15. Cantzler, H., Fisher, R.B., Devy, M.: Improving architectural 3D reconstruction by plane and edge constraining. In: Proceedings of the British Machine Vision Conference (BMVC 2002), Cardiff, U.K, pp. 43–52 (September 2002)
16. Cantzler, H., Fisher, R.B., Devy, M.: Quality enhancement of reconstructed 3D models using coplanarity and constraints. In: Van Gool, L. (ed.) DAGM 2002. LNCS, vol. 2449, pp. 34–41. Springer, Heidelberg (2002)
17. Chen, Y., Medioni, G.: Object Modelling by Registration of Multiple Range Images. In: Proceedings of the IEEE Conference on Robotics and Automation (ICRA 1991), Sacramento, CA, USA, pp. 2724–2729 (April 1991)
18. Chen, Y., Medioni, G.: Object Modelling by Registration of Multiple Range Images. Image Vision Comput. 10(3), 145–155 (1992)
19. Cole, D.M., Newman, P.M.: Using Laser Range Data for 3D SLAM in Outdoor Environments. In: Proceedings of the IEEE International Conference on Robotics and Automation (ICRA 2006), Florida, U.S.A (2006)
20. Corke, P., Cunningham, J., Dekker, D., Durrant-Whyte, H.: Autonomous underground vehicles. In: Proceedings of the CMTE Mining Technology Conference, Perth, Australia, pp. 16–22 (September 1996)
21. Cormen, T., Leiserson, C., Rivest, R.: Introduction to Algorithms. McGraw–Hill Book Company, New York (1997)
22. Dam, E., Koch, M., Lillholm, M.: Quaternion, Interpolation and Animation. Technical Report DIKU-TR-98/5, Department of Computer Science University of Copenhagen, Copenhagen, Denmark (1998)
23. DARPA (2007), `www.darpa.mil/grandchallenge/`
24. Davis, T.A.: Algorithm 849: A concise sparse Cholesky factorization package. ACM Trans. Math. Softw. 31(4) (2005)
25. Davis, T.A.: Direct Methods for Sparse Linear Systems. SIAM, Philadelphia (2006)
26. Dissanayake, M.W.M.G., Newman, P., Clark, S., Durrant-Whyte, H.F., Csorba, M.: A Solution to the Simultaneous Localization and Map Building (SLAM) Problem. IEEE Transactions on Robotics and Automation 17(3), 229–241 (2001)
27. Eggert, D., Fitzgibbon, A., Fisher, R.: Simultaneous Registration of Multiple Range Views Satisfying Global Consistency Constraints for Use In Reverse Engineering. Computer Vision and Image Understanding 69, 253–272 (1998)
28. Matrix FAQ. Version 2 (1997) `http://skal.planet-d.net/demo/matrixfaq.htm`
29. The RoboCup Federation (2007), `http://www.robocup.org/`
30. Ferguson, D., Morris, A., Hähnel, D., Baker, C., Omohundro, Z., Reverte, C., Thayer, S., Whittaker, W., Whittaker, W., Burgard, W., Thrun, S.: An autonomous robotic system for mapping abandoned mines. In: Thrun, S., Saul, L., Schölkopf, B. (eds.) Proceedings of Conference on Neural Information Processing Systems (NIPS). MIT Press, Cambridge (2003)
31. FGAN (2007), `http://www.elrob2006.org/`

32. Fischer, G.: Lineare Algebra. Friedrich Vieweg & Sohn Verglagsgesellschaft mbH, Braunschweig/Wiesbaden (1995)
33. Fitzgibbon, A.W.: Robust Registration of 2D and 3D Point Sets. In: Proceedings of the 12th British Machine Vision Conference (BMVC 2001) (2001)
34. Folkesson, J., Christensen, H.: Outdoor Exploration and SLAM using a Compressed Filter. In: Proceedings of the IEEE International Conference on Robotics and Automation (ICRA 2003), Taipei, Taiwan, pp. 419–426 (September 2003)
35. Fox, D., Thrun, S., Burgard, W., Dellaert, F.: Particle filters for mobile robot localization. In: Sequential Monte Carlo Methods in Practice. Springer, New York (2001)
36. Frehse, J.: Vorlesung Infinitisimalrechnung I – IV (Vorlesungsmitschrift). University of Bonn (1996–1997)
37. Frese, U.: Efficient 6-DOF SLAM with Treemap as a Generic Backend. In: Proceedings of the IEEE Internation Conference on Robotics and Automation (ICRA 2007), Rome, Italy (April 2007)
38. Frese, U., Hirzinger, G.: Simultaneous Localization and Mapping – A Discussion. In: Proceedings of the IJCAI Workshop on Reasoning with Uncertainty in Robotics, Seattle, USA, pp. 17–26 (August 2001)
39. Freund, Y., Schapire, R.E.: Experiments with a new boosting algorithm. In: Machine Learning: Proceedings of the 13th International Conference, pp. 148–156 (1996)
40. Friedman, J.H., Bentley, J.L., Finkel, R.A.: An algorithm for finding best matches in logarithmic expected time. ACM Transaction on Mathematical Software 3(3), 209–226 (1977)
41. Frintrop, S., Nüchter, A., Surmann, H., Hertzberg, J.: Saliency-based Object Recognition in 3D Data. In: Proceedings of the IEEE/RSJ International Conference on Intelligent Robots and Systems (IROS 2004), Sendai, Japan, pp. 2167–2172 (September 2004)
42. Früh, C., Zakhor, A.: 3D Model Generation for Cities Using Aerial Photographs and Ground Level Laser Scans. In: Proceedings of the Computer Vision and Pattern Recognition Conference (CVPR 2001), Kauai, Hawaii, USA (December 2001)
43. Früh, C., Zakhor, A.: Fast 3D Model Generation in Urban Environments. In: Proceedings of the 4th International Conference on Multisensor Fusion and Integration for Intelligent Systems (MFI 2001), Baden Baden, Germany (August 2001)
44. Fukuanaga, K., Hostetler, L.D.: Optimization of k-nearest neighbor density estimation. IEEE Transaction on Information Theory IT-19, 320–326 (1973)
45. Gelfand, N., Rusinkiewicz, S., Levoy, M.: Geometrically Stable Sampling for the ICP Algorithm. In: Proceedings of the 4th IEEE International Conference on Recent Advances in 3D Digital Imaging and Modeling (3DIM 2003), Banff, Canada, page (accepted) (October 2003)
46. Georgiev, A., Allen, P.K.: Localization Methods for a Mobile Robot in Urban Environments. IEEE Transaction on Robotics and Automation (TRO) 20(5), 851–864 (2004)
47. Giel, D.: Hologram tomography for surface topometry. PhD thesis, Mathematisch-Naturwissenschaftliche Fakultät der Heinrich-Heine- Universität Düsseldorf (2003)

48. Grau, O.: A Scene Analysis System for the Generation of 3-D Models. In: Proceedings IEEE International Conference on Recent Advances in 3D Digital Imaging and Modeling (3DIM 1997), Ottawa, Canada, pp. 221–228 (May 1997)
49. Greenspan, M., Yurick, M.: Approximate K-D Tree Search for Efficient ICP. In: Proceedings of the 4th IEEE International Conference on Recent Advances in 3D Digital Imaging and Modeling (3DIM 2003), Banff, Canada, pp. 442–448 (October 2003)
50. Greenspan, M.A., Godin, G., Talbot, J.: Acceleration of Binning Nearest Neighbor Methods. In: Proceedings of the Conference Vision Interface (2000)
51. Grisetti, G., Stachniss, C., Burgard, W.: Improved techniques for grid mapping with rao-blackwellized particle filters. IEEE Transactions on Robotics (TRO) 23(1), 34–46 (2007)
52. Guivant, J., Nebot, E., Baiker, S.: Autonomous navigation and map building using laser range sensors in outdoor applications. Journal of Robotic Systems (2000)
53. Haar, A.: Zur Theorie der orthogonalen Funktionensysteme. Mathematische Annalen (69), 331–371 (1910)
54. Hähnel, D., Burgard, W., Thrun, S.: Learning Compact 3D Models of Indoor and Outdoor Environments with a Mobile Robot. In: Proceedings of the fourth European workshop on advanced mobile robots (EUROBOT 2001), Lund, Sweden (September 2001)
55. Hamilton, W.: On a new Species of Imaginary Quantities connected with a theory of Quaternions. In: Proceedings of the Royal Irish Academy, Dublin, Ireland (November 1843)
56. Hart, J., Francis, G., Kauffmann, L.: Visualizing Quaternion Rotation. ACM Transactions on Graphics 13, 256–276 (1994)
57. Hebert, M., Deans, M., Huber, D., Nabbe, B., Vandapel, N.: Progress in 3–D Mapping and Localization. In: Proceedings of the 9th International Symposium on Intelligent Robotic Systems (SIRS 2001), Toulouse, France (July 2001)
58. Hertzberg, J., Lingemann, K., Nüchter, A.: USARSIM – Game-Engines in der Robotik-Lehre. In: Informatik 2005 – Informatik LIVE, Bonn, Germany (Beiträge der 35. Jahrestagung der Gesellschaft für Informatik, Bonn), vol. 1, pp. 158–162 (September 2005)
59. Hofer, M., Pottmann, H.: Orientierung von Laserscanner-Punktwolken. Vermessung & Geoinformation 91, 297–306 (2003)
60. Hoover, A., Jean-Baptiste, G., Jiang, X., Flynn, P.J., Bunke, H., Goldgof, D.B., Bowyer, K.K., Eggert, D.W., Fitzgibbon, A.W., Fisher, R.B.: An experimental comparison of range image segmentation algorithms. IEEE Transactions on Pattern Analysis and Machine Intelligence 18(7), 673–689 (1996)
61. Horn, B.K.P.: Closed–form solution of absolute orientation using unit quaternions. Journal of the Optical Society of America A 4(4), 629–642 (1987)
62. Horn, B.K.P., Hilden, H.M., Negahdaripour, S.: Closed–form solution of absolute orientation using orthonormal matrices. Journal of the Optical Society of America A 5(7), 1127–1135 (1988)
63. Howard, A., Roy, N. (2007), `http://radish.sourceforge.net/`
64. Huang, T., Faugeras, O.: Some Properties of the E Matrix in Two-View Motion Estimation. IEEE Transactions on Pattern Analysis and Machine Intelligence (PAMI) 11, 1310–1312 (1989)

65. Hyafil, L., Rivest, R.L.: Constructing optimal binary decision trees is np-complete. Information Processing Letters 5, 15–17 (1976)
66. Kadir, T., Zisserman, A., Brady, M.: An affine invariant salient region detector. In: Pajdla, T., Matas, J(G.) (eds.) ECCV 2004. LNCS, vol. 3021, pp. 228–241. Springer, Heidelberg (2004)
67. Katz, R., Melkumyan, N., Guivant, J., Bailey, T., Nieto, J., Nebot, E.: Integrated Sensing Framework for 3D Mapping in Outdoor Navigation. In: Proceedings of the IEEE/RSJ International Conference on Intelligent Robots and Systems (IROS 2006), Beijing, China (October 2006)
68. Kawasaki, H., Furukawa, R.: Dense 3d reconstruction method using coplanarities and metric constraints for line laser scanning. In: Godin, G., Hebert, P., Masuda, T., Taubin, G. (eds.) Proceedings of the 6th International Conference on 3D Digital Imaging and Modeling (3DIM 2007), Montreal, Canada, pp. 149–156. IEEE Computer Society Press, Los Alamitos (2007)
69. Kohlhepp, P., Walther, M., Steinhaus, P.: Schritthaltende 3D-Kartierung und Lokalisierung für mobile Inspektionsroboter. In: Dillmann, R., Wörn, H., Gockel, T. (eds.) Proceedings of the Autonome Mobile System 2003, 18. Fachgespräche (December 2003)
70. Krishnan, S., Lee, P.Y., Moore, J.B., Venkatasubramanian, S.: Global Registration of Multiple 3D Point Sets via Optimization on a Manifold. In: Eurographics Symposium on Geometry Processing (2000)
71. Kunze, L., Lingemann, K., Nüchter, A., Hertzberg, J.: Salient visual features to help close the loop in 6d slam. In: Proceedings of the ICVS Workshop on Computational Attention & Applications (WCAA 2007), Bielefeld, Germany (March 2007)
72. Langis, C., Greenspan, M., Godin, G.: The parallel iterative closest point algorithm. In: Proceedings of the 3rd IEEE International Conference on Recent Advances in 3D Digital Imaging and Modeling (3DIM 2001), Quebec City, Canada, pp. 195–204 (June 2001)
73. Lienhart, R., Kuranov, A., Pisarevsky, V.: Empirical Analysis of Detection Cascades of Boosted Classifiers for Rapid Object Detection. In: Michaelis, B., Krell, G. (eds.) DAGM 2003. LNCS, vol. 2781, pp. 297–304. Springer, Heidelberg (2003)
74. Lienhart, R., Maydt, J.: An Extended Set of Haar-like Features for Rapid Object Detection. In: Proceedings of the IEEE Conference on Image Processing (ICIP 2002), New York, USA, pp. 155–162 (Septmber 2002)
75. Lingemann, K.: Schnelles Pose-Tracking auf Laserscan-Daten für autonome mobile Roboter. Master's thesis, University of Bonn (2004)
76. Lorusso, A., Eggert, D., Fisher, R.: A Comparison of Four Algorithms for Estimating 3-D Rigid Transformations. In: Proceedings of the 4th British Machine Vision Conference (BMVC 1995), Birmingham, England, pp. 237–246 (September 1995)
77. Lu, F., Milios, E.: Robot Pose Estimation in Unknown Environments by Matching 2D Range Scans. In: IEEE Computer Vision and Pattern Recognition Conference (CVPR 1994), pp. 935–938 (1994)
78. Lu, F., Milios, E.: Globally Consistent Range Scan Alignment for Environment Mapping. Autonomous Robots 4(4), 333–349 (1997)

79. Magnusson, M., Duckett, T.: A Comparison of 3D Registration Algorithms for Autonomous Underground Mining Vehicles. In: Proceedings of the Second European Conference on Mobile Robotics (ECMR 2005), Ancona, Italy, pp. 86–91 (September 2005)
80. Matas, J., Chum, O., Martin, U., Pajdla, T.: Robust Wide Baseline Stereo from Maximally Stable Extremal Regions. In: Proceedings of the 11th British Machine Vision Conference (BMVC 2002) (2002)
81. Mikolajczyk, K., Schmid, C.: An Affine Invariant Interest Point Detector. In: Heyden, A., Sparr, G., Nielsen, M., Johansen, P. (eds.) ECCV 2002. LNCS, vol. 2350, pp. 128–142. Springer, Heidelberg (2002)
82. Mikolajczyk, K., Schmid, C.: Comparison of affine-invariant local detectors and descriptors. In: Proceedings 12th European Signal Processing Conference (2004)
83. Mikolajczyk, K., Schmid, C.: A Performance Evaluation of Local Descriptors. IEEE Transactions on Pattern Analysis and Machine Intelligence (PAMI) 27(10), 1615–1630 (2005)
84. Morris, A., Kurth, D., Whittaker, W., Thayer, S.: Case studies of a borehole deployable robot for limestone mine profiling and mapping. In: Proceedings of the International Conference on Field and Service Robotics, Lake Yamanaka, Japan (2003)
85. Nüchter, A.: Autonome Exploration und Modellierung von 3D-Umgebungen, GMD Report 157. GMD, Sankt Augustin (2002)
86. Nüchter, A.: Semantische 3D-Karten für autonome mobile Roboter (in German). PhD thesis, University of Bonn (2006)
87. Nüchter, A., Lingemann, K., Hertzberg, J., Surmann, H.: Accurate Object Localization in 3D Laser Range Scans. In: Proceedings of the 12th International Conference on Advanced Robotics (ICAR 2005), pp. 665–672 (July 2005)
88. Nüchter, A., Surmann, H., Hertzberg, J.: Automatic Classification of Objects in 3D Laser Range Scans. In: Proceedings of the 8th Conference on Intelligent Autonomous Systems (IAS 2004), Amsterdam, The Netherlands, pp. 963–970 (March 2004)
89. Nüchter, A., Surmann, H., Hertzberg, J.: Automatic Model Refinement for 3D Reconstruction with Mobile Robots. In: Proceedings of the 4th IEEE International Conference on Recent Advances in 3D Digital Imaging and Modeling (3DIM 2003), Banff, Canada, pp. 394–401 (October 2003)
90. Nüchter, A., Surmann, H., Lingemann, K., Hertzberg, J., Thrun, S.: 6D SLAM with an Application in autonomous mine mapping. In: Proceedings of the IEEE International Conference on Robotics and Automation, New Orleans, USA, pp. 1998–2003 (April 2004)
91. Nüchter, A., Lingemann, K., Hertzberg, J.: Cached k-d tree search for icp algorithms. In: Proceedings of the 6th IEEE International Conference on Recent Advances in 3D Digital Imaging and Modeling (3DIM 2007), Montreal, Canada, pp. 419–426 (August 2007)
92. NIST National Institute of Standards and Technology. Intelligent systems division (2005), `http://robotarenas.nist.gov/competitions.htm`
93. Olson, E., Leonard, J., Teller, S.: Fast iterative alignment of pose graphs with poor initial estimates. In: Proceedings of the IEEE International Conference on Robotics and Automation (2006)

94. Papageorgiou, C., Oren, M., Poggio, T.: A general framework for object detection. In: Proceedings of the 6th International Conference on Computer Vision (ICCV 1998), Bombay, India (January 1998)
95. Pauley, E., Shumaker, T., Cole, B.: Preliminary report of investigation: Underground bituminous coal mine, non-injury mine inundation accident (entrapment), July 24, Quecreek, Pennsylvania, Black Wolf Coal Company, Inc. for the PA Bureau of Deep Mine Safety (2002)
96. Pervin, E., Webb, J.: Quaternions for computer vision and robotics. In: Proceedings of the Conference on Computer Vision and Pattern Recognition (CVPR 1983), Washington, D.C, USA (1983)
97. Pfaff, P., Triebel, R., Stachniss, C., Lamon, P., Burgard, W., Siegwart, R.: Towards Mapping of Cities. In: Proceedings of the IEEE Internation Conference on Robotics and Automation (ICRA 2007), Rome, Italy (April 2007)
98. Pottmann, H., Leopoldseder, S., Hofer, M.: Simultaneous registration of multiple views of a 3D object. ISPRS Archives 34(3A), 265–270 (2002)
99. Powell, M.J.D.: An efficient method for finding the minimum of a function of several variables without calculating derivatives. Computer Journal 7, 155–162 (1964)
100. Press, W.H., Flannery, B.P., Teukolsky, S.A., Vetterling, W.T.: Numerical Recipes in C: The Art of Scientific Computing. Cambridge University Press, Cambridge (1993)
101. SWI Prolog (2003), `http://www.swi-prolog.org/`
102. Pulli, K.: Multiview Registration for Large Data Sets. In: Proceedings of the 2nd International Conference on 3D Digital Imaging and Modeling (3DIM 1999), Ottawa, Canada, pp. 160–168 (October 1999)
103. Pulli, K., Duchamp, T., Hoppe, H., McDonald, J., Shapiro, L., Stuetzle, W.: Robust meshes from multiple range maps. In: Proceedings IEEE International Conference on Recent Advances in 3D Digital Imaging and Modeling (3DIM 1997), Ottawa, Canada (May 1997)
104. RIEGL Laser Measurement Systems GmbH (2008), `www.riegl.com`
105. Rivest, R.L.: On the optimality of elias's algorithm for performing best match searches. Information Processing 74, 678–681 (1974)
106. Rocchini, C., Cignoni, P., Montani, C., Pingi, P., Scopignoy, R.: A low cost 3d scanner based on structured light. Eurographics 2001 20(3) (2001)
107. Rusinkiewicz, S., Levoy, M.: Efficient variants of the ICP algorithm. In: Proceedings of the Third International Conference on 3D Digital Imaging and Modellling (3DIM 2001), Quebec City, Canada, pp. 145–152 (May 2001)
108. Rusu, R., Marton, Z., Blodow, N., Dolha, M., Beetz, M.: Towards 3D Point Cloud Based Object Maps for Household Environments. Journal Robotics and Autonomous Systems (2008)
109. Schaffalitzky, F., Zisserman, A.: Multi-view Matching for Unordered Image Sets, or How Do I Organize My Holiday Snaps? In: Heyden, A., Sparr, G., Nielsen, M., Johansen, P. (eds.) ECCV 2002. LNCS, vol. 2350, pp. 414–431. Springer, Heidelberg (2002)
110. Scholtz, J.: Theory and evaluation of human robot interactions. In: Proceedings of the 36th Annual Hawaii International Conference on System Sciences (HICSS 2003), Washington, DC, U.S.A (2003)

111. Sequeira, V., Ng, K., Wolfart, E., Goncalves, J., Hogg, D.: Automated 3D reconstruction of interiors with multiple scan–views. In: Proceedings of SPIE, Electronic Imaging 1999, The Society for Imaging Science and Technology /SPIE's 11th Annual Symposium, San Jose, CA, USA (January 1999)
112. Shoemake, K.: Animating rotation with quaternion curves. Computer Graphics 19(3), 245–254 (1985)
113. Stachniss, C., Frese, U., Grisetti, G. (2007), http://www.openslam.org/
114. Stetson, K.A.: Holographic surface contouring by limited depth of focus. Applied Optics 7(5), 987–989 (1968)
115. Stiene, S.: Konturbasierte Objekterkennung aus Tiefenbildern eines 3D-Laserscanners. Master's thesis, University of Osnabrück (January 2006)
116. Stoddart, A., Hilton, A.: Registration of multiple point sets. In: Proceedings of the 13th IAPR International Conference on Pattern Recognition, Vienna, Austria, August 1996, pp. 40–44 (1996)
117. Surmann, H., Lingemann, K., Nüchter, A., Hertzberg, J.: A 3D laser range finder for autonomous mobile robots. In: Proceedings of the of the 32nd International Symposium on Robotics (ISR 2001), Seoul, Korea, pp. 153–158 (April 2001)
118. Surmann, H., Lingemann, K., Nüchter, A., Hertzberg, J.: Fast acquiring and analysis of three dimensional laser range data. In: Proceedings of the of the 6th International Fall Workshop Vision, Modeling, and Visualization (VMV 2001), Stuttgart, Germany, pp. 59–66 (November 2001)
119. Surmann, H., Nüchter, A., Hertzberg, J.: An autonomous mobile robot with a 3D laser range finder for 3D exploration and digitalization of indoor en vironments. Journal Robotics and Autonomous Systems 45(3–4), 181–198 (2003)
120. Thrun, S.: Robotic mapping: A survey. In: Lakemeyer, G., Nebel, B. (eds.) Exploring Artificial Intelligence in the New Millenium. Morgan Kaufmann, San Francisco (2002)
121. Thrun, S., Burgard, W., Fox, D.: A probabilistic approach to concurrent mapping and localization for mobile robots. Machine Learning and Autonomous Robots 31(5), 1–25 (1997)
122. Thrun, S., Fox, D., Burgard, W.: A real-time algorithm for mobile robot mapping with application to multi robot and 3D mapping. In: Proceedings of the IEEE International Conference on Robotics and Automation (ICRA 2000), San Francisco, CA, USA (April 2000)
123. Thrun, S., Hähnel, D., Ferguson, D., Montemerlo, M., Triebel, R., Burgard, W., Baker, C., Omohundro, Z., Thayer, S., Whittaker, W.: A system for volumetric robotic mapping of abandoned mines. In: Proceedings of the IEEE International Conference on Robotics and Automation (ICRA 2003) (2003)
124. Thrun, S., Liu, Y., Koller, D., Ng, A.Y., Ghahramani, Z., Durrant-Whyte, H.F.: Simultaneous localization and mapping with sparse extended information filters. Machine Learning and Autonomous Robots 23(7–8), 693–716 (2004)
125. Thrun, S., Montemerlo, M., Aron, A.: Probabilistic Terrain Analysis For High-Speed Desert Driving. In: Proceedings of Robotics: Science and Systems, Cambridge, USA (June 2006)
126. Torralba, A., Murphy, K.P., Freeman, W.T. (2007), http://web.mit.edu/torralba/www/database.html/

127. Triebel, R., Burgard, W.: Improving Simultaneous Localization and Mapping in 3D Using Global Constraints. In: Proceedings of the National Conference on Artificial Intelligence (AAAI 2005) (2005)
128. Triebel, R., Pfaff, P., Burgard, W.: Multi-Level Surface Maps for Outdoor Terrain Mapping and Loop Closing. In: Proceedings of the IEEE/RSJ International Conference on Intelligent Robots and Systems (IROS 2006), Beijing, China (October 2006)
129. Tuytelaars, T., Van Gool, L.: Matching Widely Separated Views Based on Affine Invariant Regions. International Journal of Computer Vision 59(1), 61–85 (2004)
130. Viola, P., Jones, M.J.: Robust real-time face detection. International Journal of Computer Vision 57(2), 137–154 (2004)
131. Völker, K. (ed.): Künstliche Menschen. Carl Hanser Verlag, München (1971)
132. Walker, M.W., Shao, L., Volz, R.A.: Estimating 3-d location parameters using dual number quaternions. CVGIP: Image Understanding 54, 358–367 (1991)
133. Weingarten, J., Siegwart, R.: EKF-based 3D SLAM for structured environment reconstruction. In: Proceedings of the IEEE/RSJ International Conference on Intelligent Robots and Systems (IROS 2005), Edmonton, Alberta Canada, pp. 2089–2094 (August 2005)
134. Williams, J.A., Bennamoun, M.: Multiple View 3D Registration using Statistical Error Models. In: Vision Modeling and Visualization (1999)
135. Winkelbach, S., Molkenstruck, S., Wahl, F.M.: Low-cost laser range scanner and fast surface registration approach. In: Franke, K., Müller, K.-R., Nickolay, B., Schäfer, R. (eds.) DAGM 2006. LNCS, vol. 4174, pp. 718–728. Springer, Heidelberg (2006)
136. Wulf, O., Arras, K.O., Christensen, H.I., Wagner, B.A.: 2D Mapping of Cluttered Indoor Environments by Means of 3D Perception. In: Proceedings of the IEEE International Conference on Robotics and Automation (ICRA 2004), New Orleans, USA, pp. 4204–4209 (April 2004)
137. Wulf, O., Brenneke, C., Wagner, B.: Colored 2D Maps for Robot Navigation with 3D Sensor Data. In: Proceedings of the IEEE International Conference on Intelligent Robots and Systems (IROS 2004), Sendai, Japan (September 2004)
138. Wulf, O., Nüchter, A., Hertzberg, J., Wagner, B.A.: Ground Truth Evaluation of Large Urban 6D SLAM. In: Proceedings of the IEEE/RSJ International Conference on Intelligent Robots and Systems (IROS 2007), San Diego, U.S.A, pp. 650–657 (October 2007)
139. Wulf, O., Wagner, B.: Fast 3D-Scanning Methods for Laser Measurement Systems. In: Proceedings of the International Conference on Control Systems and Computer Science, Bucharest, Romania (July 2003)
140. Zhang, Z.: Iterative point matching for registration of free–form curves. Technical Report RR-1658, INRIA–Sophia Antipolis, Valbonne Cedex, France (1992)
141. Zhao, H., Shibasaki, R.: Reconstructing Textured CAD Model of Urban Environment Using Vehicle-Borne Laser Range Scanners and Line Cameras. In: Second International Workshop on Computer Vision System (ICVS 2001), Vancouver, Canada, pp. 284–295 (July 2001)

Springer Tracts in Advanced Robotics

Edited by B. Siciliano, O. Khatib and F. Groen

Further volumes of this series can be found on our homepage: springer.com

Vol. 52: Nüchter, A.
3D Robotic Mapping
201 p. 2009 [978-3-540-89883-2]

Vol. 51: Song, D.
Sharing a Vision
186 p. 2009 [978-3-540-88064-6]

Vol. 50: Alterovitz, R.; Goldberg K.
Motion Planning in Medicine: Optimization and Simulation Algorithms for Image-Guided Procedures
153 p. 2008 [978-3-540-69257-7]

Vol. 49: Ott, C.
Cartesian Impedance Control of Redundant and Flexible-Joint Robots
190 p. 2008 [978-3-540-69253-9]

Vol. 48: Wolter, D.
Spatial Representation and Reasoning for Robot Mapping
185 p. 2008 [978-3-540-69011-5]

Vol. 47: Akella, S.; Amato, N.; Huang, W.; Mishra, B.; (Eds.)
Algorithmic Foundation of Robotics VII
524 p. 2008 [978-3-540-68404-6]

Vol. 46: Bessière, P.; Laugier, C.; Siegwart R. (Eds.)
Probabilistic Reasoning and Decision Making in Sensory-Motor Systems
375 p. 2008 [978-3-540-79006-8]

Vol. 45: Bicchi, A.; Buss, M.; Ernst, M.O.; Peer A. (Eds.)
The Sense of Touch and Its Rendering
281 p. 2008 [978-3-540-79034-1]

Vol. 44: Bruyninckx, H.; Přeučil, L.; Kulich, M. (Eds.)
European Robotics Symposium 2008
356 p. 2008 [978-3-540-78315-2]

Vol. 43: Lamon, P.
3D-Position Tracking and Control for All-Terrain Robots
105 p. 2008 [978-3-540-78286-5]

Vol. 42: Laugier, C.; Siegwart, R. (Eds.)
Field and Service Robotics
597 p. 2008 [978-3-540-75403-9]

Vol. 41: Milford, M.J.
Robot Navigation from Nature
194 p. 2008 [978-3-540-77519-5]

Vol. 40: Birglen, L.; Laliberté, T.; Gosselin, C.
Underactuated Robotic Hands
241 p. 2008 [978-3-540-77458-7]

Vol. 39: Khatib, O.; Kumar, V.; Rus, D. (Eds.)
Experimental Robotics
563 p. 2008 [978-3-540-77456-3]

Vol. 38: Jefferies, M.E.; Yeap, W.-K. (Eds.)
Robotics and Cognitive Approaches to Spatial Mapping
328 p. 2008 [978-3-540-75386-5]

Vol. 37: Ollero, A.; Maza, I. (Eds.)
Multiple Heterogeneous Unmanned Aerial Vehicles
233 p. 2007 [978-3-540-73957-9]

Vol. 36: Buehler, M.; Iagnemma, K.; Singh, S. (Eds.)
The 2005 DARPA Grand Challenge – The Great Robot Race
520 p. 2007 [978-3-540-73428-4]

Vol. 35: Laugier, C.; Chatila, R. (Eds.)
Autonomous Navigation in Dynamic Environments
169 p. 2007 [978-3-540-73421-5]

Vol. 34: Wisse, M.; van der Linde, R.Q.
Delft Pneumatic Bipeds
136 p. 2007 [978-3-540-72807-8]

Vol. 33: Kong, X.; Gosselin, C.
Type Synthesis of Parallel Mechanisms
272 p. 2007 [978-3-540-71989-2]

Vol. 32: Milutinović, D.; Lima, P.
Cells and Robots – Modeling and Control of Large-Size Agent Populations
130 p. 2007 [978-3-540-71981-6]

Vol. 31: Ferre, M.; Buss, M.; Aracil, R.; Melchiorri, C.; Balaguer C. (Eds.)
Advances in Telerobotics
500 p. 2007 [978-3-540-71363-0]

Vol. 30: Brugali, D. (Ed.)
Software Engineering for Experimental Robotics
490 p. 2007 [978-3-540-68949-2]

Vol. 29: Secchi, C.; Stramigioli, S.; Fantuzzi, C.
Control of Interactive Robotic Interfaces – A Port-Hamiltonian Approach
225 p. 2007 [978-3-540-49712-7]

Vol. 28: Thrun, S.; Brooks, R.; Durrant-Whyte, H. (Eds.)
Robotics Research – Results of the 12th International Symposium ISRR
602 p. 2007 [978-3-540-48110-2]

Vol. 27: Montemerlo, M.; Thrun, S.
FastSLAM – A Scalable Method for the Simultaneous Localization and Mapping Problem in Robotics
120 p. 2007 [978-3-540-46399-3]

Vol. 26: Taylor, G.; Kleeman, L.
Visual Perception and Robotic Manipulation – 3D Object Recognition, Tracking and Hand-Eye Coordination
218 p. 2007 [978-3-540-33454-5]

Vol. 25: Corke, P.; Sukkarieh, S. (Eds.)
Field and Service Robotics – Results of the 5th International Conference
580 p. 2006 [978-3-540-33452-1]

Vol. 24: Yuta, S.; Asama, H.; Thrun, S.; Prassler, E.; Tsubouchi, T. (Eds.)
Field and Service Robotics – Recent Advances in Research and Applications
550 p. 2006 [978-3-540-32801-8]

Vol. 23: Andrade-Cetto, J,; Sanfeliu, A.
Environment Learning for Indoor Mobile Robots – A Stochastic State Estimation Approach to Simultaneous Localization and Map Building
130 p. 2006 [978-3-540-32795-0]

Vol. 22: Christensen, H.I. (Ed.)
European Robotics Symposium 2006
209 p. 2006 [978-3-540-32688-5]

Vol. 21: Ang Jr., H.; Khatib, O. (Eds.)
Experimental Robotics IX – The 9th International Symposium on Experimental Robotics
618 p. 2006 [978-3-540-28816-9]

Vol. 20: Xu, Y.; Ou, Y.
Control of Single Wheel Robots
188 p. 2005 [978-3-540-28184-9]

Vol. 19: Lefebvre, T.; Bruyninckx, H.; De Schutter, J. Nonlinear Kalman Filtering for Force-Controlled Robot Tasks
280 p. 2005 [978-3-540-28023-1]

Vol. 18: Barbagli, F.; Prattichizzo, D.; Salisbury, K. (Eds.)
Multi-point Interaction with Real and Virtual Objects
281 p. 2005 [978-3-540-26036-3]

Vol. 17: Erdmann, M.; Hsu, D.; Overmars, M.; van der Stappen, F.A (Eds.)
Algorithmic Foundations of Robotics VI
472 p. 2005 [978-3-540-25728-8]

Vol. 16: Cuesta, F.; Ollero, A.
Intelligent Mobile Robot Navigation
224 p. 2005 [978-3-540-23956-7]

Vol. 15: Dario, P.; Chatila R. (Eds.)
Robotics Research – The Eleventh International Symposium
595 p. 2005 [978-3-540-23214-8]

Vol. 14: Prassler, E.; Lawitzky, G.; Stopp, A.; Grunwald, G.; Hägele, M.; Dillmann, R.; Iossifidis. I. (Eds.)
Advances in Human-Robot Interaction
414 p. 2005 [978-3-540-23211-7]

Vol. 13: Chung, W.
Nonholonomic Manipulators
115 p. 2004 [978-3-540-22108-1]

Vol. 12: Iagnemma K.; Dubowsky, S.
Mobile Robots in Rough Terrain – Estimation, Motion Planning, and Control with Application to Planetary Rovers
123 p. 2004 [978-3-540-21968-2]

Vol. 11: Kim, J.-H.; Kim, D.-H.; Kim, Y.-J.; Seow, K.-T.
Soccer Robotics
353 p. 2004 [978-3-540-21859-3]

Vol. 10: Siciliano, B.; De Luca, A.; Melchiorri, C.; Casalino, G. (Eds.)
Advances in Control of Articulated and Mobile Robots
259 p. 2004 [978-3-540-20783-2]

Vol. 9: Yamane, K.
Simulating and Generating Motions of Human Figures
176 p. 2004 [978-3-540-20317-9]

Vol. 8: Baeten, J.; De Schutter, J.
Integrated Visual Servoing and Force Control – The Task Frame Approach
198 p. 2004 [978-3-540-40475-0]

Vol. 7: Boissonnat, J.-D.; Burdick, J.; Goldberg, K.; Hutchinson, S. (Eds.)
Algorithmic Foundations of Robotics V
577 p. 2004 [978-3-540-40476-7]